T0214629

# Lecture Notes in Mathematics 2239

More information about this series at http://www.springer.com/series/304

Mauro Di Nasso • Isaac Goldbring • Martino Lupini

# Nonstandard Methods in Ramsey Theory and Combinatorial Number Theory

Springer

Mauro Di Nasso
Department of Mathematics
Universita di Pisa
Pisa, Italy

Isaac Goldbring
Department of Mathematics
University of California, Irvine
Irvine, CA, USA

Martino Lupini
School of Mathematics and Statistics
Victoria University of Wellington
Wellington, New Zealand

ISSN 0075-8434                    ISSN 1617-9692    (electronic)
Lecture Notes in Mathematics
ISBN 978-3-030-17955-7           ISBN 978-3-030-17956-4    (eBook)
https://doi.org/10.1007/978-3-030-17956-4

Mathematics Subject Classification (2010): Primary: 05D10, Secondary: 03H10

This Springer imprint is published by the registered company Springer Nature Switzerland AG
The registered company address is: Gewerbestrasse 11, 6330 Cham, Switzerland

*To Marian, Karina, and Nicola*

# Preface

Generally speaking, Ramsey theory studies which combinatorial configurations of a structure can always be found in one of the pieces of a given finite partition. More generally, it considers the problem of which combinatorial configurations can be found in sets that are "large" in some suitable sense. Dating back to the foundational results of van der Waerden, Ramsey, Erdős, Turán, and others from the 1920s and 1930s, Ramsey theory has since then had an extraordinary development. On the one hand, many applications of Ramsey theory to numerous other areas of mathematics, ranging from functional analysis, topology, and dynamics to set theory, model theory, and computer science, have been found. On the other hand, results and methods from other areas of mathematics have been successfully applied to establish new results in Ramsey theory. For instance, ergodic theory has had a profound impact on Ramsey theory, giving rise to the research area of "ergodic Ramsey theory." Perhaps the best known achievement of this approach is the ergodic-theoretic proof of Szemerédi's theorem due to Furstenberg in the 1980s. In a different (but intimately related) direction, the theory of ultrafilters has been an important source of methods and ideas for Ramsey theory. In particular, the study of topological and algebraic properties of the space of ultrafilters has been used to give short and elegant proofs of deep combinatorial pigeonhole principles. Paradigmatic in this direction is the Galvin–Glazer ultrafilter proof of Hindman's theorem on sets of finite sums, previously established by Hindman in 1974 via a delicate, purely combinatorial argument.

Recently, a new thread of research has emerged, where problems in Ramsey theory are studied from the perspective of nonstandard analysis and nonstandard methods. Developed by Abraham Robinson in the 1960s and based on first-order logic and model theory, nonstandard analysis provided a formal and rigorous treatment of calculus and classical analysis via infinitesimals. Such a treatment is more similar in spirit to the approach originally taken in the development of calculus in the seventeenth and eighteenth century and avoids the epsilon-delta arguments that are inherent in its later formalization due to Weierstrass. While this is perhaps its most well-known application, nonstandard analysis is actually much more versatile. The foundations of nonstandard analysis provide us with a method, which we shall call the nonstandard method, that is applicable to virtually any area of mathematics. The nonstandard method has thus far been used in numerous areas of mathematics, including functional analysis, measure theory, ergodic theory, differential equations, and stochastic analysis, just to name a few such areas.

In a nutshell, the nonstandard method allows one to extend the given mathematical universe and thus regard it as contained in a much richer nonstandard universe. Such a nonstandard universe satisfies strong saturation properties which in particular allow one to consider limiting objects which do not exist in the standard universe. This procedure is similar to passing to an ultrapower, and in fact the nonstandard method can also be seen as a way to axiomatize the ultrapower construction in a way that distillates its essential features and benefits, but avoids being bogged down by the irrelevant details of its concrete implementation. This limiting process allows one to reformulate a given problem involving finite (but arbitrarily large) structures or configurations into a problem involving a single structure or configuration which is infinite but for all purposes behaves as though it were finite (in the precise sense that it is hyperfinite in the nonstandard universe). This reformulation can then be tackled directly using finitary methods, ranging from combinatorial counting arguments to recurrence theorems for measurable dynamics, recast in the nonstandard universe.

In the setting of Ramsey theory and combinatorics, the application of nonstandard methods was pioneered by the work of Keisler, Leth, and Jin from the 1980s and 1990s. These applications focused on density problems in combinatorial number theory. The general goal in this area is to establish the existence of combinatorial configurations in sets that are large in the sense that they have positive asymptotic density. For example, the aforementioned celebrated theorem of Szemerédi from 1970 asserts that a set of integers of positive density contains arbitrarily long finite arithmetic progressions. One of the contributions of the nonstandard approach is to translate the notion of asymptotic density on the integers, which does not satisfy all the properties of a measure, into an actual measure in the nonstandard universe. This translation then makes methods from measure theory and ergodic theory, such as the ergodic theorem or other recurrence theorems, available for the study of density problems. In a sense, this can be seen as a version of Furstenberg's correspondence (between sets of integers and measurable sets in a dynamical system), with the extra feature that the dynamical system obtained perfectly reflects *all* the combinatorial properties of the set that one started with. The achievements of the nonstandard approach in this area include the work of Leth on arithmetic progressions in sparse sets, Jin's theorem on sumsets, as well as Jin's Freiman-type results on inverse problems for sumsets. More recently, these methods have also been used by Jin, Leth, Mahlburg, and the present authors to tackle a conjecture of Erdős concerning sums of infinite sets (the so-called $B + C$ conjecture), leading to its eventual solution by Moreira, Richter, and Robertson.

Nonstandard methods are also tightly connected with ultrafilter methods. This has been made precise and successfully applied in a recent work of one of us (Di Nasso), where he observed that there is a perfect correspondence between ultrafilters and elements of the nonstandard universe up to a natural notion of equivalence. On the one hand, this allows one to manipulate ultrafilters as nonstandard points and to use ultrafilter methods to prove the existence of certain combinatorial configurations in the nonstandard universe. On the other hand, this gives an intuitive and direct way to infer, from the existence of certain ultrafilter configurations, the existence of corresponding standard combinatorial configurations via the fundamental principle of transfer from nonstandard analysis. This perspective has successfully been applied by Di Nasso and Luperi Baglini to the study of partition regularity problems for Diophantine equations over the integers, providing in particular a far-reaching generalization of the classical theorem of Rado on partition regularity of systems of linear equations. Unlike Rado's theorem, this recent generalization also includes equations that are *not* linear.

Finally, it is worth mentioning that many other results in combinatorics can be seen, directly or indirectly, as applications of the nonstandard method. For instance, the groundbreaking work of Hrushovski and Breuillard–Green–Tao on approximate groups, although not originally presented in this way, admit a natural nonstandard treatment. The same applies to the work of Bergelson and Tao on recurrence in quasirandom groups.

The goal of this manuscript is to introduce the uninitiated reader to the nonstandard method and to provide an overview of its most prominent applications in Ramsey theory and combinatorial number theory. In particular, no previous knowledge of nonstandard analysis will be assumed. Instead, we will provide a complete and self-contained introduction to the nonstandard method in the first part of this book. Novel to our introduction is a treatment of the topic of iterated hyperextensions, which is crucial for some applications and has thus far appeared only in specialized research articles. The intended audience for this book includes researchers in combinatorics who desire to get acquainted with the nonstandard approach, as well as logicians and experts of nonstandard analysis who have been working in this or other areas of research. The list of applications of the nonstandard method to combinatorics and Ramsey theory presented here is quite extensive, including cornerstone results of Ramsey theory such as Ramsey's theorem, Hindman's theorem on sets of finite sums, the Hales–Jewett theorem on variable words, and Gowers' Ramsey theorem. It then proceeds with results on partition regularity of Diophantine equations and with density problems in combinatorial number theory. A nonstandard treatment of the triangle removal lemma, the Szemerédi regularity lemma, and of the already mentioned work of Hrushovski and Breuillard–Green–Tao on approximate groups conclude the book. We hope that such a complete list of examples will help the reader unfamiliar with the nonstandard method get a good grasp on how the approach works and can be applied. At the same time, we believe that collecting these results together, and providing a unified presentation and approach, will provide a useful reference for researchers in the field and will further stimulate the research in this currently very active area.

This work originated out of lecture notes written by the authors on occasion of the workshop "Nonstandard methods in combinatorial number theory" held in 2017 at the American Institute of Mathematics. We would like to thank the participants of the conference for providing valuable feedback to an earlier version of the manuscript. We received further useful comments and suggestions from several people, including Kevin Barnum, Jordan Barrett, Ryan Burkhart, the Bogotá Logic Group (especially Alex Berenstein and Dario Garcia), and Lorenzo Luperi Baglini. We would also like to express our gratitude to our friend and colleague Steven Leth for his many helpful insights on earlier versions of this draft (as well as for being responsible for bringing the second author into this line of research).

Di Nasso's research was partially supported by PRIN Grant 2012 "Models and Set Theory." Goldbring's work was partially supported by NSF CAREER grant DMS-1349399. Lupini was partially supported by the NSF Grant DMS-1600186 and by a Research Establishment Grant by Victoria University of Wellington.

Pisa, Italy                                                                  Mauro Di Nasso
Irvine, CA, USA                                                          Isaac Goldbring
Wellington, New Zealand                                        Martino Lupini

# Notation and Conventions

We set $\mathbb{N} := \{1, 2, 3, \ldots\}$ to denote the set of *strictly positive* natural numbers and $\mathbb{N}_0 := \{0, 1, 2, 3, \ldots\}$ to denote the set of natural numbers.

We use the following conventions for elements of particular sets:

- $m$ and $n$ range over $\mathbb{N}$.
- $h, k$, and $l$ range over $\mathbb{Z}$.
- $H, K, M$, and $N$ range over elements of $^*\mathbb{N}$.
- $\delta$ and $\epsilon$ sometimes denote (small) positive elements of $\mathbb{R}$, while sometimes they denote positive infinitesimal elements of $^*\mathbb{R}$. There will be no confusion between these two uses as we will always explicitly state the assumptions on the quantity.
- Given any set $S$, we let $\alpha, \beta$, and $\gamma$ denote arbitrary (possibly standard) elements of $^*S$.

For any set $S$, we set $\mathrm{Fin}(S) := \{F \subseteq S \mid F \text{ is finite and nonempty}\}$.

For any $n \in \mathbb{N}$, we write $[n] := \{1, \ldots, n\}$. Similarly, for $N \in {}^*\mathbb{N}$ we write $[N] := \{1, \ldots, N\}$.

Given any nonempty finite set $I$ and any set $A$, we write $\delta(A, I) := \frac{|A \cap I|}{|I|}$. We extend this to the nonstandard situation: if $I$ is a nonempty hyperfinite set and $A$ is an internal set, we set $\delta(A, I) := \frac{|A \cap I|}{|I|}$. We also write $\delta(A, n) := \delta(A, [n])$ and $\delta(A, N) := \delta(A, [N])$.

Given a hyperfinite set $X$, we let $\mathscr{L}_X$ denote the $\sigma$-algebra of Loeb measurable subsets of $X$ and we let $\mu_X$ denote the Loeb measure on $\mathscr{L}_X$ that extends the normalized counting measure on $X$ (see Chap. 5). When $X = \{1, \ldots, N\}$, we write $\mathscr{L}_N$ and $\mu_N$ instead of $\mathscr{L}_X$ and $\mu_X$. If $A$ is internal, we write $\mu_X(A) := \mu_X(A \cap X)$.

Suppose that $A \subseteq \mathbb{Z}$ and $k \in \mathbb{N}$. We write

$$k \cdot A := \{x_1 + \cdots + x_k : x_1, \ldots, x_k \in A\}$$

and

$$kA := \{kx : x \in A\}.$$

Of course $kA \subseteq k \cdot A$.

Throughout this book, log always denotes the logarithm base 2.

# Contents

## Part III   Combinatorial Number Theory

## Part IV   Other Topics

# Part I
# Preliminaries

# Chapter 1
# Ultrafilters

Filters and ultrafilters—initially introduced by Cartan in the 1930s—are fundamental objects in mathematics and naturally arise in many different contexts. They are featured prominently in Bourbaki's systematic treatment of general topology as they allow one to capture the central notion of "limit". The general definition of limit of a filter subsumes, in particular, the usual limit of a sequence (or net). Ultrafilters—a special class of filters—are especially useful, as they always admit a limit provided they are defined on a compact space. Thus, ultrafilters allow one to extend the usual notion of limit to sequences (or nets) that would not have a limit in the usual sense. Such "generalized limits" are frequently used in asymptotic or limiting arguments, where compactness is used in a fundamental way.

## 1.1 Basics on Ultrafilters

Throughout this chapter, we let $S$ denote an infinite set.

**Definition 1.1** A (proper) *filter* on $S$ is a set $\mathscr{F}$ of subsets of $S$ (that is, $\mathscr{F} \subseteq \mathscr{P}(S)$) such that:

- $\emptyset \notin \mathscr{F}$, $S \in \mathscr{F}$;
- if $A, B \in \mathscr{F}$, then $A \cap B \in \mathscr{F}$;
- if $A \in \mathscr{F}$ and $A \subseteq B$, then $B \in \mathscr{F}$.

We think of elements of $\mathscr{F}$ as "big" sets (because that is what filters do, they catch the big objects). The first and third axioms are (hopefully) intuitive properties of big sets. Perhaps the second axiom is not as intuitive, but if one thinks of the complement of a big set as a "small" set, then the second axiom asserts that the union of two small sets is small (which is hopefully more intuitive).

© Springer Nature Switzerland AG 2019
M. Di Nasso et al., *Nonstandard Methods in Ramsey Theory
and Combinatorial Number Theory*, Lecture Notes in Mathematics 2239,
https://doi.org/10.1007/978-3-030-17956-4_1

**Exercise 1.2** Set $\mathscr{F} := \{A \subseteq S \mid S \backslash A$ is finite$\}$. Prove that $\mathscr{F}$ is a filter on $S$, called the *Frechét* or *cofinite* filter on $S$.

**Definition 1.3** Suppose that $\mathscr{D}$ is a set of subsets of $S$. We say that $\mathscr{D}$ has the *finite intersection property* if: whenever $D_1, \ldots, D_n \in \mathscr{D}$, we have $D_1 \cap \cdots \cap D_n \neq \emptyset$.

**Exercise 1.4** Suppose that $\mathscr{D}$ is a set of subsets of $S$ with the finite intersection property. Set

$$\langle \mathscr{D} \rangle := \{E \subseteq S \mid D_1 \cap \cdots \cap D_n \subseteq E \text{ for some } D_1, \ldots, D_n \in \mathscr{D}\}.$$

Show that $\langle \mathscr{D} \rangle$ is the smallest filter on $S$ containing $\mathscr{D}$, called the *filter generated by $\mathscr{D}$*.

If $\mathscr{F}$ is a filter on $S$, then a subset of $S$ cannot be simultaneously big and small (that is, both it and its complement belong to $\mathscr{F}$), but there is no requirement that it be one of the two. It will be desirable (for reasons that will become clear in a moment) to add this as an additional property:

**Definition 1.5** If $\mathscr{F}$ is a filter on $S$, then $\mathscr{F}$ is an *ultrafilter* if, for any $A \subseteq S$, either $A \in \mathscr{F}$ or $S \backslash A \in \mathscr{F}$ (but not both!).

Ultrafilters are usually denoted by $\mathscr{U}$. Observe that the Frechét filter on $S$ is not an ultrafilter since there are sets $A \subseteq S$ such that $A$ and $S \backslash A$ are both infinite.

The following exercise illustrates one of the most important properties of ultrafilters .

**Exercise 1.6** A filter $\mathscr{F}$ on $S$ is an ultrafilter if and only if whenever $A_1, \ldots, A_n$ are subsets of $S$ such that $A_1 \cup \cdots \cup A_n \in \mathscr{F}$, there exists $i \in \{1, \ldots, n\}$ such that $A_i \in \mathscr{F}$.

We have yet to see an example of an ultrafilter. Here is a "trivial" source of ultrafilters:

**Definition 1.7** Given $s \in S$, set $\mathscr{U}_s := \{A \subseteq S \mid s \in A\}$.

**Exercise 1.8** For $s \in S$, prove that $\mathscr{U}_s$ is an ultrafilter on $S$, called the *principal ultrafilter generated by $s$*.

We say that an ultrafilter $\mathscr{U}$ on $S$ is *principal* if $\mathscr{U} = \mathscr{U}_s$ for some $s \in S$. Although principal ultrafilters settle the question of the existence of ultrafilters, they will turn out to be useless for our purposes, as we will soon see. From a philosophical viewpoint, principal ultrafilters fail to capture the idea that sets belonging to the ultrafilter are large, for $\{s\}$ belongs to the ultrafilter $\mathscr{U}_s$ and yet hardly anyone would dare say that the set $\{s\}$ is large!

**Exercise 1.9** Prove that an ultrafilter $\mathscr{U}$ on $S$ is nonprincipal if and only if there is no *finite* set $A \subseteq S$ such that $A \in \mathscr{U}$ if and only if $\mathscr{U}$ extends the Frechét filter.

We now would like to prove the existence of nonprincipal ultrafilters. The following exercise will be the key to doing this.

**Exercise 1.10** Suppose that $\mathscr{F}$ is a filter on $S$. Then $\mathscr{F}$ is an ultrafilter on $S$ if and only if it is a maximal filter, that is, if and only if, whenever $\mathscr{F}'$ is a filter on $S$ such that $\mathscr{F} \subseteq \mathscr{F}'$, we have $\mathscr{F} = \mathscr{F}'$.

Since it is readily verified that the union of an increasing chain of filters on $S$ containing a filter $\mathscr{F}$ is once again a filter on $S$ containing $\mathscr{F}$, the previous exercises and Zorn's lemma yield the following:

**Theorem 1.11** *Nonprincipal ultrafilters on $S$ exist.*

**Exercise 1.12** Suppose that $f : S \to T$ is a function between sets. Then given any ultrafilter $\mathscr{U}$ on $S$, the set

$$f(\mathscr{U}) := \{A \subseteq T \,:\, f^{-1}(A) \in \mathscr{U}\}$$

is an ultrafilter on $T$, called the *image ultrafilter of $\mathscr{U}$ under $f$*.

## 1.2   The Space of Ultrafilters $\beta S$

In this section, $S$ continues to denote an infinite set. Since topological matters are the subject of this subsection, we will also treat $S$ as a topological space equipped with the discrete topology.

The set of ultrafilters on $S$ is denoted $\beta S$. There is a natural topology on $\beta S$ obtained by declaring, for $A \subseteq S$, the following sets as basic open sets:

$$U_A := \{\mathscr{U} \in \beta S \,:\, A \in \mathscr{U}\}.$$

(Note that the $U_A$'s are indeed a base for a topology as $U_A \cap U_B = U_{A \cap B}$.) Since the complement of $U_A$ in $\beta S$ is $U_{S \setminus A}$, we see that the basic open sets are in fact clopen. Note also that $\beta S$ is Hausdorff: if $\mathscr{U}, \mathscr{V} \in \beta S$ are distinct, take $A \subseteq S$ with $A \in \mathscr{U}$ and $S \setminus A \in \mathscr{V}$. Then $\mathscr{U} \in U_A$ and $\mathscr{V} \in U_{S \setminus A}$ and $U_A$ and $U_{S \setminus A}$ are disjoint.

**Exercise 1.13** Let $\mathscr{A}$ be a family of subsets of $S$. Then the following (possibly empty) set is closed in $\beta S$:

$$\mathscr{C}_{\mathscr{A}} := \{\mathscr{U} \in \beta S \,:\, \mathscr{A} \subseteq \mathscr{U}\}.$$

**Theorem 1.14** $\beta S$ *is a compact space.*

*Proof* It is enough to show that every family $(C_i)$ with the intersection property and consisting of compact sets has nonempty intersection. Without loss of generality, we can assume that $C_i = U_{A_i}$ for some subsets $A_i$ of $S$. The assumption that the family $(U_{A_i})$ has the finite intersection property implies that the family $(A_i)$ has the finite intersection property. Thus there exists an ultrafilter $\mathscr{U}$ over $S$ such that $A_i \in \mathscr{U}$ for every $i$. Thus $\mathscr{U} \in U_{A_i}$ for every $i$, witnessing that $(C_i)$ has nonempty intersection.

We identify $S$ with the set of principal ultrafilters on $S$. Under this identification, $S$ is dense in $\beta S$: if $A \subseteq S$ is nonempty and $s \in A$, then the principal ultrafilter $\mathcal{U}_s \in U_A$. Thus, $\beta S$ is a compactification of $S$. In fact, we have:

**Theorem 1.15** $\beta S$ *is the Stone-Čech compactification of* $S$.

We remind the reader that the Stone-Čech compactification of $S$ is the unique compactification $X$ of $S$ with the following property: any function $f : S \to Y$ with $Y$ compact Hausdorff has a unique continuous extension $\tilde{f} : X \to Y$. In order to prove the previous theorem, we will first need the following lemma, which is important in its own right:

**Lemma 1.16** *Suppose that* $Y$ *is a compact Hausdorff space and* $(y_s)_{s \in S}$ *is a family of elements of* $Y$ *indexed by* $S$. *Then for any* $\mathcal{U} \in \beta S$, *there is a unique element* $y \in Y$ *with the property that, for any open neighborhood* $U$ *of* $y$, *we have* $\{s \in S : y_s \in U\} \in \mathcal{U}$.

*Proof* Suppose, towards a contradiction, that no such $y$ exists. Then for every $y \in Y$, there is an open neighborhood $U_y$ of $y$ such that $\{s \in S : y_s \in U_y\} \notin \mathcal{U}$. By compactness, there are $y_1, \ldots, y_n \in Y$ such that $Y = U_{y_1} \cup \cdots \cup U_{y_n}$. There exists then $i \in \{1, \ldots, n\}$ such that $\{s \in S : y_s \in U_{y_i}\} \in \mathcal{U}$, yielding the desired contradiction. We now show that such a $y$ is unique. Suppose by contradiction that there exists $y' \neq y$ satisfying the same conclusions. Then as by assumption $Y$ is Hausdorff, one can choose disjoint open neighborhoods $U_y$ and $U_{y'}$ of $y$ and $y'$, respectively. Hence, we have that $\mathcal{U}$ contains both $\{s \in S : y_s \in U_y\}$ and $\{s \in S : y_s \in U_{y'}\}$. This contradicts the assumption that $\mathcal{U}$ is an ultrafilter, as these two sets are disjoint.

**Definition 1.17** In the context of the previous lemma, we call the unique $y$ the *ultralimit of* $(y_s)$ *with respect to* $\mathcal{U}$, denoted $\lim_{s, \mathcal{U}} y_s$ or simply just $\lim_{\mathcal{U}} y_s$.

*Proof (of Theorem 1.15)* Suppose that $f : S \to Y$ is a function into a compact Hausdorff space. Define $\tilde{f} : \beta S \to Y$ by $\tilde{f}(\mathcal{U}) := \lim_{\mathcal{U}} f(s)$, which exists by Lemma 1.16. It is clear that $\tilde{f}(\mathcal{U}_s) = f(s)$, so $\tilde{f}$ extends $f$. We must show that $\tilde{f}$ is continuous. Fix $\mathcal{U} \in \beta S$ and let $U$ be an open neighborhood of $\tilde{f}(\mathcal{U})$ in $Y$. Let $V \subseteq U$ be an open neighborhood of $\tilde{f}(\mathcal{U})$ in $Y$ such that $\overline{V} \subseteq U$. (This is possible because every compact Hausdorff space is *regular*, that is, every point has a base of closed neighborhoods.) Take $A \in \mathcal{U}$ such that $f(s) \in V$ for $s \in A$. Suppose $\mathcal{V} \in U_A$, so $A \in \mathcal{V}$. Then $\lim_{\mathcal{V}} f(s) \in \overline{V} \subseteq U$, so $U_A \subseteq \tilde{f}^{-1}(U)$.

Recall that the set of points where two continuous functions with values in a Hausdroff space agree is closed. As $S$ is dense in $\beta S$, it follows that such a continuous extension is unique.

Now that we have shown that $\beta S$ is the Stone-Čech compactification of $S$, given $f : S \to Y$ where $Y$ is a compact Hausdorff space, we will let $\beta f : \beta S \to Y$ denote the unique continuous extension of $f$. If $S, T$ are two sets and $f : S \to T$ is a function, then we let $\beta f : \beta S \to \beta T$ be the continuous extension of $f$.

**Definition 1.18** Fix $k \in \mathbb{N}$. Let $m_k : \mathbb{N} \to \mathbb{N}$ be defined by $m_k(n) := kn$. Then for $\mathcal{U} \in \beta\mathbb{N}$, we set $k\mathcal{U} := (\beta m_k)(\mathcal{U})$.

Note that $A \in k\mathcal{U} \Leftrightarrow A/k := \{n \in \mathbb{N} \mid nk \in A\} \in \mathcal{U}$. The ultrafilters $k\mathcal{U}$ will play an important role in Chap. 9.

**Exercise 1.19** Given $A \subseteq S$, show that $\overline{A} = U_A$, where $\overline{A}$ denotes the closure of $A$ in $\beta S$.

Let $\ell^\infty(S)$ denote the space of bounded real-valued functions on $S$. Given $f \in \ell^\infty(S)$, take $r \in \mathbb{R}^{>0}$ such that $f(S) \subseteq [-r, r]$, whence we may consider its unique continuous extension $\beta f : \beta S \to [-r, r]$. Note that the function $\beta f$ does not depend on the choice of $r$. The following exercise will be useful in Chap. 12.

**Exercise 1.20** The function $f \mapsto \beta f$ is an isomorphism between $\ell^\infty(S)$ and $C(\beta S)$ as Banach spaces.

## 1.3 The Case of a Semigroup

We now suppose that $S$ is the underlying set of a semigroup $(S, \cdot)$. Then one can extend the semigroup operation $\cdot$ to a semigroup operation $\odot$ on $\beta S$ by declaring, for $\mathcal{U}, \mathcal{V} \in \beta S$ and $A \subseteq S$, that

$$A \in \mathcal{U} \odot \mathcal{V} \Leftrightarrow \{s \in S : s^{-1} \cdot A \in \mathcal{V}\} \in \mathcal{U}.$$

Here, $s^{-1}A := \{t \in S : s \cdot t \in A\}$. In other words, $\mathcal{U} \odot \mathcal{V} = \lim_{s, \mathcal{U}} (\lim_{t, \mathcal{V}} s \cdot t)$, where these limits are taken in the compact space $\beta S$. In particular, note that $\mathcal{U}_s \odot \mathcal{U}_t = \mathcal{U}_{s \cdot t}$, so this operation on $\beta S$ does indeed extend the original operation on $S$. It is also important to note that, in general, ultralimits do not commute and thus, in general, $\mathcal{U} \odot \mathcal{V} \neq \mathcal{V} \odot \mathcal{U}$, even if $(S, \cdot)$ is commutative. (See Chap. 3 for more on this lack of commutativity.) In the commutative case, we often denote the semigroup operation by $+$, in which case we write the extended operation on the space of ultrafilters by $\oplus$.

The following theorem allows one to apply Theorem 1.23 below to produce ultrafilters whose existence has deep combinatorial consequences.

**Theorem 1.21** $(\beta S, \odot)$ *is a compact,* right topological semigroup, *that is, $\odot$ is a semigroup operation on the compact space $\beta S$ such that, for each $\mathcal{V} \in \beta S$, the left translation map $\rho_\mathcal{V} : \mathcal{U} \mapsto \mathcal{U} \odot \mathcal{V} : \beta S \to \beta S$ is continuous.*

*Proof* Fix $\mathcal{V} \in \beta S$. We need to show that $\rho_\mathcal{V}$ is continuous. Towards this end, fix $A \subseteq S$. We must show that $\rho_\mathcal{V}^{-1}(U_A)$ is open. Let $B := \{s \in S : s^{-1} \cdot A \in \mathcal{V}\}$. It remains to note that $\rho_\mathcal{V}^{-1}(U_A) = U_B$.

One can introduce some notation to express more succinctly the semigroup operation in $\beta S$. Given $A \subseteq S$ and $t \in S$, one defines $A \cdot t^{-1}$ to be the set $\{s \in S : s \cdot t \in A\}$. Similarly, for $A \subseteq S$ and $\mathcal{V} \in \beta S$, one defines $A \cdot \mathcal{V}^{-1}$

to be $\{x \in S : x^{-1} \cdot A \in \mathcal{V}\}$. Then for $\mathcal{U}, \mathcal{V} \in \beta S$ and $A \subseteq S$, one has that $A \in \mathcal{U} \odot \mathcal{V}$ if and only if $A \cdot \mathcal{V}^{-1} \in \mathcal{U}$.

## 1.4  The Existence of Idempotents in Semitopological Semigroups

**Definition 1.22** Suppose that $(S, \cdot)$ is a semigroup. We say that $e \in S$ is *idempotent* if $e \cdot e = e$.

The following classical theorem of Ellis is the key to many applications of ultrafilter methods and nonstandard methods in Ramsey theory.

**Theorem 1.23** *Suppose that $(S, \cdot)$ is a compact semitopological semigroup. Then $S$ has an idempotent element.*

*Proof* Let $\mathscr{S}$ denote the set of nonempty closed subsemigroups of $S$. It is clear that the intersection of any descending chain of elements of $\mathscr{S}$ is also an element of $\mathscr{S}$, whence by Zorn's lemma, we may find $T \in \mathscr{S}$ that is minimal.

Fix $s \in T$. We show that $s$ is idempotent. Set $T_1 := Ts$. Note that $T_1 \neq \emptyset$ as $T \neq \emptyset$. Since $S$ is a semitopological semigroup and $T$ is compact, we have that $T_1$ is also compact. Finally, note that $T_1$ is also a subsemigroup of $S$:

$$T_1 \cdot T_1 = (Ts)(Ts) \subseteq T \cdot T \cdot T \cdot s \subseteq T \cdot s = T_1.$$

We thus have that $T_1 \in \mathscr{S}$. Since $s \in T$, we have that $T_1 \subseteq T$, whence by minimality of $T$, we have that $T_1 = T$. In particular, the set $T_2 := \{t \in T : t \cdot s = s\}$ is not empty. Note that $T_2$ is also a closed subset of $T$, whence compact. Once again, we note that $T_2$ is a subsemigroup of $S$. Indeed, if $t, t' \in T_2$, then $tt' \in T$ and $(tt') \cdot s = t \cdot (t' \cdot s) = t \cdot s = s$. We thus have that $T_2 \in \mathscr{S}$. By minimality of $T$, we have that $T_2 = T$. It follows that $s \in T_2$, that is, $s \cdot s = s$.

The previous theorem and Theorem 1.21 immediately give the following:

**Corollary 1.24** *Let $(S, \cdot)$ be a semigroup and let $T$ be any nonempty closed subsemigroup of $(\beta S, \odot)$. Then $T$ contains an idempotent element.*

We refer to idempotent elements of $\beta S$ as *idempotent ultrafilters*. Thus, the previous corollary says that any nonempty closed subsemigroup of $\beta S$ contains an idempotent ultrafilter.

## 1.5  Partial Semigroups

We will encounter the need to apply the ideas in the previous section to the broader context of partial semigroups.

**Definition 1.25**  A *partial semigroup* is a set $S$ endowed with a partially defined binary operation $(s, t) \mapsto s \cdot t$ that satisfies the following form of the associative law: given $s_1, s_2, s_3 \in S$, if either of the products $(s_1 \cdot s_2) \cdot s_3$ or $s_1 \cdot (s_2 \cdot s_3)$ are defined, then so is the other and the products are equal. The partial semigroup $(S, \cdot)$ is *directed* if, for any finite subset $F$ of $S$, there exists $t \in S$ such that the product $s \cdot t$ is defined for every $s \in F$.

The following example will play a central role in our discussion of Gowers' theorem in Chap. 8.

*Example 1.26*  For $k \in \mathbb{N}$, we let $\text{FIN}_k$ denote the set of functions $b : \mathbb{N} \to \{0, 1, \ldots, k\}$ with $\text{Supp}(b)$ finite and such that $k$ belongs to the range of $b$. Here, $\text{Supp}(b) := \{n \in \mathbb{N} : b(n) \neq 0\}$ is the *support of $b$*. Note that, after identifying a subset of $\mathbb{N}$ with its characteristic function, $\text{FIN}_1$ is simply the set of nonempty finite subsets of $\mathbb{N}$. We endow $\text{FIN}_k$ with a partial semigroup operation $(b_0, b_1) \mapsto b_0 + b_1$, which is defined only when every element of $\text{Supp}(b_0)$ is less than every element of $\text{Supp}(b_1)$. It is clear that this partial semigroup is in fact directed.

For the rest of this section, we assume that $(S, \cdot)$ is a directed partial semigroup.

**Definition 1.27**  We call $\mathcal{U} \in \beta S$ *cofinite* if, for all $s \in S$, we have $\{t \in S : s \cdot t \text{ is defined}\} \in \mathcal{U}$. We let $\gamma S$ denote the set of all cofinite elements of $\beta S$.

**Exercise 1.28**  $\gamma S$ is a *nonempty* closed subset of $\beta S$.

We can define an operation $\odot$ on $\gamma S$ by declaring, for $\mathcal{U}, \mathcal{V} \subset \gamma S$ and $A \subseteq S$, that $A \in \mathcal{U} \odot \mathcal{V}$ if and only if

$$\{s \in S : \{t \in S : s \cdot t \text{ is defined and } s \cdot t \in A\} \in \mathcal{V}\} \in \mathcal{U}.$$

Note that the operation $\odot$ is a totally defined operation on $\gamma S$ even though the original operation $\cdot$ was only a partially defined operation.

**Exercise 1.29**  $(\gamma S, \odot)$ is a compact semitopological semigroup. Consequently, every nonempty closed subsemigroup of $\gamma S$ contains an idempotent element.

## Notes and References

The notion of ultrafilter was introduced by Cartan [23, 24] in 1937 to study convergence in topological spaces. Ultrafilters and the corresponding construction of ultraproduct are common tools in mathematical logic, but they also found many applications in other fields of mathematics, especially in topology, algebra, and functional analysis. A classic reference on ultrafilters is the book *"Ultrafilters"* by Comfort and Negrepontis [30]. See also the more recent [14] for a review of ultrafilters across mathematics. The standard reference for the algebra of $\beta S$ and its applications throughout Ramsey theory is the excellent book by Hindman and Strauss [72]. The extension of the operation on a semigroup to the space of ultrafilters can be seen as a particular instance of the notion of Arens product on the bidual of a Banach algebra [2]. Indeed, one can regard the space of ultrafilters over a semigroup $S$ as a subspace of the second dual of the Banach algebra $\ell^1(S)$ endowed with the usual convolution product. This was the initial approach taken in the study of the Stone-Čech compactification since the 1950s [29, 33]. Its realization as a space of ultrafilters was first explicitly considered by Ellis [44]. The existence of idempotent elements in any compact right topological semigroup is a classical result of Ellis [44]. The observation that this implies the existence of idempotent ultrafilters is due to Galvin. Idempotent ultrafilters play a fundamental role in the application of ultrafilter methods to combinatorics, starting from the Galvin–Glazer proof of Hindman's Theorem on sumsets (see Chap. 8 below).

# Chapter 2
# Nonstandard Analysis

If one wants to present the methods of nonstandard analysis in their full generality and with full rigor, then notions and tools from mathematical logic such as "first-order formula" or "elementary extension" are definitely needed. However, we believe that a gentle introduction to the basics of nonstandard methods and their use in combinatorics does not directly require any technical machinery from logic. Only at a later stage, when advanced nonstandard techniques are applied and their use must be put on firm foundations, detailed knowledge of notions from logic will be necessary.

We will begin with presenting the main properties of the nonstandard versions of the natural, integer, rational, and real numbers, which will be named by adding the prefix "hyper". Then we will introduce the fundamental principle of nonstandard analysis, namely the *transfer principle* of the *star map*. While at this stage the treatment will still be informal, it will still be sufficient for the reader to gain a first idea of how nonstandard methods can be used in applications.

In the Appendix, we give sound and rigorous foundations to nonstandard analysis in full generality by formally introducing first order logic. The reader already familiar with nonstandard methods can proceed directly to the next chapter.

## 2.1 Warming-Up

To begin with, let us recall the following notions, which are at the very base of nonstandard analysis.

**Definition 2.1** An element $\varepsilon$ of an ordered field $\mathbb{F}$ is *infinitesimal* (or *infinitely small*) if $-\frac{1}{n} < \varepsilon < \frac{1}{n}$ for every $n \in \mathbb{N}$. A number $\Omega$ is *infinite* if either $\Omega > n$ for every $n \in \mathbb{N}$ or $\Omega < -n$ for every $n \in \mathbb{N}$.

© Springer Nature Switzerland AG 2019
M. Di Nasso et al., *Nonstandard Methods in Ramsey Theory
and Combinatorial Number Theory*, Lecture Notes in Mathematics 2239,
https://doi.org/10.1007/978-3-030-17956-4_2

In Definition 2.1 we identify a natural number $n$ with the element of $\mathbb{F}$ obtained as the $n$-fold sum of 1 by itself. Clearly, a nonzero number is infinite if and only if its reciprocal is infinitesimal. We say that a number is *finite* or *bounded* if it is not infinite.

**Exercise 2.2**

1. If $\xi$ and $\zeta$ are finite, then $\xi + \zeta$ and $\xi \cdot \zeta$ are finite.
2. If $\xi$ and $\zeta$ are infinitesimal, then $\xi + \zeta$ is infinitesimal.
3. If $\xi$ is infinitesimal and $\zeta$ is finite, then $\xi \cdot \zeta$ is infinitesimal.
4. If $\xi$ is infinite and $\zeta$ is not infinitesimal, then $\xi \cdot \zeta$ is infinite.
5. If $\xi$ is infinitesimal and $\zeta$ is not infinitesimal, then $\xi / \zeta$ is infinitesimal.
6. If $\xi$ is infinite and $\zeta$ is finite, then $\xi / \zeta$ is infinite.

Recall that an ordered field $\mathbb{F}$ is *Archimedean* if for every positive $x \in \mathbb{F}$ there exists $n \in \mathbb{N}$ such that $nx > 1$.

**Exercise 2.3** The following properties are equivalent for an ordered field $\mathbb{F}$:

1. $\mathbb{F}$ is non-Archimedean;
2. There are nonzero infinitesimal numbers in $\mathbb{F}$;
3. The set of natural numbers has an upper bound in $\mathbb{F}$.

We are now ready to introduce the nonstandard reals.

**Definition 2.4** The *hyperreal field* $^*\mathbb{R}$ is a proper extension of the ordered field $\mathbb{R}$ that satisfies additional properties (to be specified further on). The element of $^*\mathbb{R}$ are called *hyperreal numbers*.

By just using the above incomplete definition, the following is proved.

**Proposition 2.5** *The hyperreal field $^*\mathbb{R}$ is non-Archimedean, and hence it contains nonzero infinitesimals and infinite numbers.*

*Proof* Since $^*\mathbb{R}$ is a proper extension of the real field, we can pick a number $\xi \in {}^*\mathbb{R} \backslash \mathbb{R}$. Without loss of generality, let us assume $\xi > 0$. If $\xi$ is infinite, then we are done. Otherwise, by the completeness property of $\mathbb{R}$, we can consider the number $r = \inf\{x \in \mathbb{R} \mid x > \xi\}$. (Notice that it may be $r < \xi$.) It is readily checked that $\xi - r$ is a nonzero infinitesimal number.

We remark that, as a non-Archimedean field, $^*\mathbb{R}$ is *not* complete (e.g., the set of infinitesimals is bounded but has no least upper bound). We say that two hyperreal numbers are *infinitely close* if their difference is infinitesimal. The nonstandard counterpart of completeness is given by the following property.

**Theorem 2.6 (Standard Part)** *Every finite hyperreal number $\xi \in {}^*\mathbb{R}$ is infinitely close to a unique real number $r \in \mathbb{R}$, called the* standard part *of $\xi$. In this case, we use the notation $r = \mathrm{st}(\xi)$.*

*Proof*  By the completeness of $\mathbb{R}$, we can set $\text{st}(\xi) := \inf\{x \in \mathbb{R} \mid x > \xi\} = \sup\{y \in \mathbb{R} \mid y < \xi\}$. By the supremum (or infimum) property, it directly follows that $\text{st}(\xi)$ is infinitely close to $\xi$. Moreover, $\text{st}(\xi)$ is the unique real number with that property, since infinitely close real numbers are necessarily equal.

It follows that every finite hyperreal number $\xi$ has a unique representation in the form $\xi = r + \varepsilon$ where $r = \text{st}(\xi) \in \mathbb{R}$ and $\varepsilon$ is infinitesimal. By definition of standard part, $\xi \in {}^*\mathbb{R}$ is infinitesimal if and only if $\text{st}(\xi) = 0$. Given finite hyperreals $\xi$ and $\zeta$, it is sometimes convenient to write $\xi \gtrsim \zeta$ to mean $\text{st}(\xi) \geq \text{st}(\zeta)$.

The following are the counterparts in the nonstandard setting of the familiar properties of limits of real sequences.

**Exercise 2.7**  For all finite hyperreal numbers $\xi, \zeta$:

1. $\text{st}(\xi) < \text{st}(\zeta) \Rightarrow \xi < \zeta \Rightarrow \text{st}(\xi) \leq \text{st}(\zeta)$;
2. $\text{st}(\xi + \zeta) = \text{st}(\xi) + \text{st}(\zeta)$;
3. $\text{st}(\xi \cdot \zeta) = \text{st}(\xi) \cdot \text{st}(\zeta)$;
4. $\text{st}(\frac{\xi}{\zeta}) = \frac{\text{st}(\xi)}{\text{st}(\zeta)}$ whenever $\zeta$ is not infinitesimal.

**Definition 2.8**  The *ring of hyperinteger numbers* ${}^*\mathbb{Z}$ is an unbounded discretely ordered subring of ${}^*\mathbb{R}$ that satisfies special properties (to be specified further on), including the following:

- For every $\xi \in {}^*\mathbb{R}$ there exists $\zeta \in {}^*\mathbb{Z}$ with $\zeta \leq \xi < \zeta + 1$. Such a $\zeta$ is called the *hyperinteger part* of $\xi$, denoted $\zeta = \lfloor \xi \rfloor$.

Since ${}^*\mathbb{Z}$ is discretely ordered, notice that its finite part coincides with $\mathbb{Z}$. This means that for every $z \in \mathbb{Z}$ there are *no* hyperintegers $\zeta \in {}^*\mathbb{Z}$ such that $z < \zeta < z + 1$.

**Definition 2.9**  The *hypernatural numbers* ${}^*\mathbb{N}$ are the positive part of ${}^*\mathbb{Z}$. Thus ${}^*\mathbb{Z} = -{}^*\mathbb{N} \cup \{0\} \cup {}^*\mathbb{N}$, where $-{}^*\mathbb{N} = \{-\xi \mid \xi \in {}^*\mathbb{N}\}$ are the negative hyperintegers.

**Definition 2.10**  The field of *hyperrational numbers* ${}^*\mathbb{Q}$ is the subfield of ${}^*\mathbb{R}$ consisting of elements of the form $\frac{\xi}{\nu}$ where $\xi \in {}^*\mathbb{Z}$ and $\nu \in {}^*\mathbb{N}$.

**Exercise 2.11**  The hyperrational numbers ${}^*\mathbb{Q}$ are dense in ${}^*\mathbb{R}$, that is, for every pair $\xi < \xi'$ in ${}^*\mathbb{R}$ there exists $\eta \in {}^*\mathbb{Q}$ such that $\xi < \eta < \xi'$.

We remark that, although still incomplete, our definitions suffice to get a clear picture of the order-structure of the two main nonstandard objects that we will consider here, namely the hypernatural numbers ${}^*\mathbb{N}$ and the hyperreal line ${}^*\mathbb{R}$. In particular, let us focus on the nonstandard natural numbers. One possible way (but certainly not the only possible way) to visualize them is the following:

- The *hypernatural numbers* ${}^*\mathbb{N}$ are the extended version of the natural numbers that is obtained by allowing the use of a "mental telescope" to also see infinite numbers beyond the finite ones.

So, beyond the usual finite numbers $\mathbb{N} = \{1, 2, 3, \ldots\}$, one finds infinite numbers $\xi > n$ for all $n \in \mathbb{N}$. Every $\xi \in {}^*\mathbb{N}$ has a successor $\xi + 1$, and every non-zero $\xi \in {}^*\mathbb{N}$ has a predecessor $\xi - 1$.

$$
{}^*\mathbb{N} = \Big\{ \underbrace{1, 2, 3, \ldots, n, \ldots}_{\text{finite numbers}} \quad \underbrace{\ldots, N-2, N-1, N, N+1, N+2, \ldots}_{\text{infinite numbers}} \Big\}
$$

Thus the set of finite numbers $\mathbb{N}$ does not have a greatest element and the set of infinite numbers ${}^*\mathbb{N}\backslash\mathbb{N}$ does not have a least element, whence ${}^*\mathbb{N}$ is *not* well-ordered. Sometimes we will write $\nu > \mathbb{N}$ to mean that $\nu \in {}^*\mathbb{N}$ is infinite.

**Exercise 2.12** Consider the equivalence relation $\sim_f$ on ${}^*\mathbb{N}$ defined by setting $\xi \sim_f \zeta$ if $\xi - \zeta$ is finite. The corresponding equivalence classes are called *galaxies*. The quotient set ${}^*\mathbb{N}/\!\sim_f$ inherits an order structure, which turns it into a dense linearly ordered set with least element $[1] = \mathbb{N}$ and with no greatest element.

## 2.2   The Star Map and the Transfer Principle

As we have seen in the previous section, corresponding to each of the sets $\mathbb{N}, \mathbb{Z}, \mathbb{Q}, \mathbb{R}$, one has a *nonstandard extension*, namely the sets ${}^*\mathbb{N}, {}^*\mathbb{Z}, {}^*\mathbb{Q}, {}^*\mathbb{R}$, respectively. A defining feature of nonstandard analysis is that one has a canonical way of extending *every* mathematical object $A$ under study to an object ${}^*A$ which inherits all "elementary" properties of the initial object.

**Definition 2.13** The *star map* is a function that associates to each "mathematical object" $A$ under study its *hyper-extension* (or *nonstandard extension*) ${}^*A$ in such a way that the following holds:

- *Transfer principle:* Let $P(A_1, \ldots, A_n)$ be an "elementary property" of the mathematical objects $A_1, \ldots, A_n$. Then $P(A_1, \ldots, A_n)$ is true if and only if $P({}^*A_1, \ldots, {}^*A_n)$ is true:

$$
P(A_1, \ldots, A_n) \iff P({}^*A_1, \ldots, {}^*A_n).
$$

One can think of hyper-extensions as a sort of weakly isomorphic copy of the initial objects. Indeed, by the *transfer principle*, an object $A$ and its hyper-extension ${}^*A$ are indistinguishable as far as their "elementary properties" are concerned. Of course, the crucial point here is to precisely determine which properties are "elementary" and which are not.

Let us remark that the above definition is indeed incomplete in that the notions of "mathematical object" and of "elementary property" are still to be made precise and rigorous. As anticipated in the introduction, we will do this gradually.

To begin with, it will be enough to include in our notion of "mathematical object" the following:

1. Real numbers and $k$-tuples of real numbers for every $k \in \mathbb{N}$;
2. All sets $A \subseteq \mathbb{R}^k$ of real tuples, and all functions $f : A \to B$ between them;
3. All sets made up of objects in (1) and (2), including, e.g., the families $\mathscr{F} \subseteq \bigcup_k \mathscr{P}(\mathbb{R}^k)$ of sets of real $k$-tuples, and the families of functions $\mathscr{G} \subseteq \mathrm{Fun}(\mathbb{R}^k, \mathbb{R}^h)$.

More generally, every other structure under study could be safely included in the list of "mathematical objects".[1]

As for the notion of "elementary property", we will start working with a semi-formal definition. Although not fully rigorous from a logical point of view, it may nevertheless look perfectly fine to many, and we believe that it can be safely adopted to get introduced to nonstandard analysis and to familiarize oneself with its basic notions and tools.

**Definition 2.14** A property $P$ is *elementary* if it can be expressed by an *elementary formula*, that is, by a formula where:

1. Besides the usual logical connectives ("not", "and", "or", "if ... then", "if and only if") and the quantifiers ("there exists", "for every") only the basic notions of equality, membership, set, ordered $k$-tuple, $k$-ary relation, domain, range, function, value of a function at a given point, are involved;
2. The scope of every quantifier is *bounded*, that is, quantifiers always occur in the form "there exists $x \in X$" or "for every $y \in Y$" for specified sets $X, Y$. More generally, also nested quantifiers "$Q\,x_1 \in x_2$ and $Q\,x_2 \in x_3 \ldots$ and $Q\,x_n \in X$" are allowed, where $Q$ is either "there exists" or "for every", $x_1, \ldots, x_n$ are variables, and $X$ is a specified set.

An immediate consequence of the *transfer* principle is that all fundamental mathematical constructions are preserved under the star map, with the only two relevant exceptions being *powersets* and *function sets* (see Proposition 2.49). Below we give three comprehensive lists in three distinct propositions, the first one about sets and ordered tuples, the second one about relations, and the third one about functions. Until the notion of "elementary property" has been made precise, one can take those properties as axioms for the star map.

**Proposition 2.15**

1. $a = b \Leftrightarrow {}^*a = {}^*b$.
2. $a \in A \Leftrightarrow {}^*a \in {}^*A$.
3. $A$ is a set if and only if ${}^*A$ is a set.

---

[1]According to the usual set-theoretic foundational framework, every mathematical object is identified with a set (see Remark A.2 in the Appendix). However, here we will stick to the common perception that considers numbers, ordered pairs, relations, functions, and sets as mathematical objects of distinct nature.

4. $^*\emptyset = \emptyset$.

If $A, A_1, \ldots, A_k, B$ are sets:

5. $A \subseteq B \Leftrightarrow {}^*A \subseteq {}^*B$.
6. $^*(A \cup B) = {}^*A \cup {}^*B$.
7. $^*(A \cap B) = {}^*A \cap {}^*B$.
8. $^*(A \backslash B) = {}^*A \backslash {}^*B$.
9. $^*\{a_1, \ldots, a_k\} = \{{}^*a_1, \ldots, {}^*a_k\}$.
10. $^*(a_1, \ldots, a_k) = ({}^*a_1, \ldots, {}^*a_k)$.
11. $^*(A_1 \times \ldots \times A_k) = {}^*A_1 \times \ldots \times {}^*A_k$.
12. $^*\{(a, a) \mid a \in A\} = \{(\xi, \xi) \mid \xi \in {}^*A\}$.

If $\mathscr{F}$ is a family of sets:

13. $^*\{(x, y) \mid x \in y \in \mathscr{F}\} = \{(\xi, \zeta) \mid \xi \in \zeta \in {}^*\mathscr{F}\}$.
14. $^*(\bigcup_{F \in \mathscr{F}} F) = \bigcup_{G \in {}^*\mathscr{F}} G$.

Proof  Recall that by our definition, the notions of equality, membership, set, and ordered $k$-tuple are elementary; thus by direct applications of *transfer* one obtains (1), (2), (3), and (10), respectively. All other properties are easily proved by considering suitable elementary formulas. As examples, we will consider here only three of them.

(8) The property "$C = A \backslash B$" is elementary, because it is formalized by the elementary formula:

$$\text{"}\forall x \in C \ (x \in A \text{ and } x \notin B) \text{ and } \forall x \in A \ (x \notin B \Rightarrow x \in C)\text{"}.$$

So, by *transfer*, we have that $C = A \backslash B$ holds if and only if

$$\text{"}\forall x \in {}^*C \ (x \in {}^*A \text{ and } x \notin {}^*B) \text{ and } \forall x \in {}^*A \ (x \notin {}^*B \Rightarrow x \in {}^*C)\text{"},$$

that is, if and only if $^*C = {}^*A \backslash {}^*B$.

(9) The property "$C = \{a_1, \ldots, a_k\}$" is formalized by the elementary formula: "$a_1 \in C$ and $\ldots\ldots$ and $a_k \in C$ and $\forall x \in C \ (x = a_1 \text{ or } \ldots \text{ or } x = a_k)$". So, we can apply *transfer* and obtain that $^*C = \{{}^*a_1, \ldots, {}^*a_k\}$.

(14) The property "$A = \bigcup_{F \in \mathscr{F}} F$" is formalized by the elementary formula: "$\forall x \in A \ (\exists y \in \mathscr{F} \text{ with } x \in y)$ and $\forall y \in \mathscr{F} \ \forall x \in y \ (x \in A)$." Then by *transfer* one gets "$^*A = \bigcup_{y \in {}^*\mathscr{F}} y$."

**Proposition 2.16**

1. $R$ is a $k$-ary relation if and only if $^*R$ is a $k$-ary relation.

If $R$ is a binary relation:

2. $^*\{a \mid \exists b \ R(a, b)\} = \{\xi \mid \exists \zeta \ {}^*R(\xi, \zeta)\}$, that is, $^*domain(R) = domain({}^*R)$.
3. $^*\{b \mid \exists a \ R(a, b)\} = \{\zeta \mid \exists \xi \ {}^*R(\xi, \zeta)\}$, that is, $^*range(R) = range({}^*R)$.
4. $^*\{(a, b) \mid R(b, a)\} = \{(\xi, \zeta) \mid {}^*R(\zeta, \xi)\}$.

*If S is a ternary relation:*

5. $^*\{(a, b, c) \mid S(c, a, b)\} = \{(\xi, \zeta, \eta) \mid {}^*S(\eta, \xi, \zeta)\}$.
6. $^*\{(a, b, c) \mid S(a, c, b)\} = \{(\xi, \zeta, \eta) \mid {}^*S(\xi, \eta, \zeta)\}$.

*Proof* (1), (2), and (3) are proved by direct applications of *transfer*, because the notions of $k$-ary relation, domain, and range are elementary by definition.

(4). The property "$C = \{(a, b) \mid R(b, a)\}$" is formalized by the conjunction of the elementary formula "$\forall z \in C \; \exists x \in$ domain$(R) \; \exists y \in$ range$(R)$ s.t. $R(x, y)$ and $z = (y, x)$" and the elementary formula "$\forall x \in$ domain$(R) \; \forall y \in$ range$(R) \; (y, x) \in C$". Thus *transfer* applies and one obtains $^*C = \{(\xi, \zeta) \mid (\zeta, \xi) \in {}^*R\}$.

(5) and (6) are proved by considering similar elementary formulas as in (4).

**Proposition 2.17**

1. $f$ *is a function if and only if* $^*f$ *is a function.*

   *If* $f, g$ *are functions and* $A, B$ *are sets:*

2. $^*domain(f) = domain(^*f)$.
3. $^*range(f) = range(^*f)$.
4. $f : A \to B$ *if and only if* $^*f : {}^*A \to {}^*B$.[2]
5. $^*graph(f) = graph(^*f)$.
6. $^*(f(a)) = (^*f)(^*a)$ *for every* $a \in domain(f)$.
7. *If* $f : A \to A$ *is the identity, then* $^*f : {}^*A \to {}^*A$ *is the identity, that is* $^*(1_A) = 1_{^*A}$.
8. $^*\{f(a) \mid a \in A\} = \{^*f(\xi) \mid \xi \in {}^*A\}$, *that is* $^*(f(A)) = {}^*f(^*A)$.
9. $^*\{a \mid f(a) \in B\} = \{\xi \mid {}^*f(\xi) \in {}^*B\}$, *that is* $^*(f^{-1}(B)) = (^*f)^{-1}(^*B)$.
10. $^*(f \circ g) = {}^*f \circ {}^*g$.
11. $^*\{(a, b) \in A \times B \mid f(a) = g(b)\} = \{(\xi, \zeta) \in {}^*A \times {}^*B \mid {}^*f(\xi) = {}^*g(\zeta)\}$.

*Proof* (1), (2), (3), and (6) are proved by direct applications of *transfer*, because the notions of function, value of a function at a given point, domain, and range, are elementary. (4) is a direct corollary of the previous properties. We only prove two of the remaining properties as all of the proofs are similar to one another.

(5). The property "$C = graph(f)$" is formalized by the elementary formula obtained as the conjunction of the formula "$\forall z \in C \; \exists x \in$ domain$(f) \; \exists y \in$ range$(f)$ such that $y = f(x)$ and $(x, y) \in C$" with the formula "$\forall x \in$ domain$(f) \; \forall y \in$ range$(f) \; (y = f(x) \Rightarrow (x, y) \in C)$". The desired equality follows by *transfer* and by the previous properties.

(10). If $f : A \to B$ and $g : B \to C$, then the property "$h = g \circ f$" is formalized by the formula "$h : A \to C$ and $\forall x \in A \; \forall y \in C \; (h(x) = y \Leftrightarrow \exists z \in B \; f(x) = z$ and $g(z) = y)$".

---

[2]Recall that notation $f : A \to B$ means that $f$ is a function with domain$(f) = A$ and range$(f) \subseteq B$.

**Exercise 2.18** Prove that a function $f : A \to B$ is 1-1 if and only if $^*f : {}^*A \to {}^*B$ is 1-1.

We now discuss a general result about the star map that is really useful in practice (and, in fact, several particular cases have already been included in the previous propositions): If a set is defined by means of an elementary property, then its hyper-extension is defined by the same property where one puts stars in front of the parameters.

**Proposition 2.19** *Let $\varphi(x, y_1, \ldots, y_n)$ be an elementary formula. For all objects $B, A_1, \ldots, A_n$ one has*

$$^*\{x \in B \mid \varphi(x, A_1, \ldots, A_n)\} = \{x \in {}^*B \mid \varphi(x, {}^*A_1, \ldots, {}^*A_n)\}.$$

*Proof* Since $C = \{x \in B \mid \varphi(x, A_1, \ldots, A_n)\}$, we have that the following formula holds:

$$P(A_1, \ldots, A_n, B, C) : \forall x \, (x \in C \Leftrightarrow (x \in B \text{ and } \varphi(x, A_1, \ldots, A_n))).$$

By *transfer*, we have that $P({}^*A_1, \ldots, {}^*A_n, {}^*B, {}^*C)$ holds as well. This readily implies that is $^*C = \{x \in {}^*B \mid \varphi(x, {}^*A_1, \ldots, {}^*A_n)\}$.

An immediate corollary is the following.

**Proposition 2.20** *If $(a, b) = \{x \in \mathbb{R} \mid a < x < b\}$ is an open interval of real numbers then $^*(a, b) = \{\xi \in {}^*\mathbb{R} \mid a < \xi < b\}$, and similarly for intervals of the form $[a, b), (a, b], (a, b), (-\infty, b]$ and $[a, +\infty)$. Analogous properties hold for intervals of natural, integer, or rational numbers.*

## 2.2.1   Additional Assumptions

By property Proposition 2.15 (1) and (2), the hyper-extension $^*A$ of a set $A$ contains a copy of $A$ given by the hyper-extensions of its elements

$$^\sigma A = \{{}^*a \mid a \in A\} \subseteq {}^*A.$$

Notice that, by *transfer*, an hyper-extension $^*x$ belongs to $^*A$ if and only if $x \in A$. Therefore, $^\sigma A \cap {}^\sigma B = {}^\sigma(A \cap B)$ for all sets $A, B$.

Following the common use in nonstandard analysis, to simplify matters we will assume that $^*r = r$ for all $r \in \mathbb{R}$. This implies by transfer that $^*(r_1, \ldots, r_k) = (r_1, \ldots, r_k)$ for all ordered tuples of real numbers, i.e. $^\sigma(\mathbb{R}^k) = \mathbb{R}^k$. It follows that hyper-extensions of real sets and functions are actual extensions:

- $A \subseteq {}^*A$ for every $A \subseteq \mathbb{R}^k$,
- If $f : A \to B$ where $A \subseteq \mathbb{R}^k$ and $B \subseteq \mathbb{R}^h$, then $^*f$ is an extension of $f$, that is, $^*f(a) = f(a)$ for every $a \in A$.

In nonstandard analysis it is always assumed that the star map satisfies the following

- *Properness condition:* $^*\mathbb{N} \neq \mathbb{N}$.

**Proposition 2.21** *If the properness condition* $^*\mathbb{N} \neq \mathbb{N}$ *holds, then* $^\sigma A \neq {}^*A$ *for every infinite A.*

*Proof* Suppose that there is an infinite set $A$ such that $^\sigma A = {}^*A$. Fix a surjective map $f : A \to \mathbb{N}$. Then also the hyper-extension $^*f : {}^*A \to {}^*\mathbb{N}$ is surjective, and

$$^*\mathbb{N} = \{^*f(\alpha) \mid \alpha \in {}^*A\} = \{^*f(^*a) \mid a \in A\} = \{^*(f(a)) \mid a \in A\}$$
$$= \{^*n \mid n \in \mathbb{N}\} = \mathbb{N},$$

whence the properness condition fails. □

As a first consequence of the properness condition, one gets a nonstandard characterization of finite sets as those sets that are not "extended" by hyper-extensions.

**Proposition 2.22** *For every set A one has the equivalence: "A is finite if and only if* $^*A = {}^\sigma A$". *(When* $A \subseteq \mathbb{R}^k$, *this is the same as "A is finite if and only if* $^*A = A$".)*

*Proof* If $A = \{a_1, \ldots, a_k\}$ is finite, we already saw in Proposition 2.15 (9) that $^*A = \{^*a_1, \ldots, {}^*a_k\} = \{^*a \mid a \in A\}$. Conversely, if $A$ is infinite, we can pick a surjective function $f : A \to \mathbb{N}$. Then also $^*f : {}^*A \to {}^*\mathbb{N}$ is onto. Now notice that for every $a \in A$, one has that $(^*f)(^*a) = {}^*(f(a)) = f(a) \in \mathbb{N}$ (recall that $^*n = n$ for every $n \in \mathbb{N}$). Then if $\xi \in {}^*\mathbb{N}\backslash\mathbb{N}$ there exists $\alpha \in {}^*A\backslash\{^*a \mid a \in A\}$ with $^*f(\alpha) = \xi$. □

One can safely extend the simplifying assumption $^*r = r$ from real numbers $r$ to elements of any given mathematical object $X$ under study (but of course not for all mathematical objects).

- *Unless explicitly mentioned otherwise, when studying a specific mathematical object X by nonstandard analysis, we will assume that* $^*x = x$ *for all* $x \in X$, *so that* $X = {}^\sigma X \subseteq {}^*X$.

It is worth mentioning at this point that star maps satisfying the *transfer principle* and the *properness condition* do actually exist. Indeed, they can be easily constructed by means of ultrafilters, or, equivalently, by means of maximal ideals of rings of functions (see Sect. 2.4).

We end this section with an example of using properness to give a short proof of a classical fact. (This nonstandard proof was suggested by D.A. Ross.)

**Theorem 2.23 (Sierpinski)** *Given* $a_1, \ldots, a_n, b \in \mathbb{R}^{>0}$, *the set*

$$E := \left\{ (x_1, \ldots, x_n) \in \mathbb{N}^n : \frac{a_1}{x_1} + \cdots + \frac{a_n}{x_n} = b \right\}$$

*is finite.*

*Proof* Suppose, towards a contradiction, that $E$ is infinite. Then there is $x = (x_1, \ldots, x_n) \in {}^*E \backslash E$. Without loss of generality, we may assume that there is $k \in \{1, \ldots, n\}$ such that $x_1, \ldots, x_k \in {}^*\mathbb{N} \backslash \mathbb{N}$ and $x_{k+1}, \ldots, x_n \in \mathbb{N}$. We then have

$$\frac{a_1}{x_1} + \cdots + \frac{a_k}{x_k} = b - \left( \frac{a_{k+1}}{x_{k+1}} + \cdots + \frac{a_n}{x_n} \right).$$

We have now arrived at a contradiction for the left hand side of the equation is a positive infinitesimal element of ${}^*\mathbb{R}$ while the right hand side of the equation is a positive standard real number.

## 2.3  The Transfer Principle, in Practice

As we already pointed out, a correct application of *transfer* needs a precise understanding of the notion of elementary property. Basically, a property is elementary if it talks about the elements of some given structures and *not* about their subsets or the functions between them.[3] Indeed, in order to correctly apply the *transfer principle*, one must always point out the range of quantifiers, and formulate them in the forms "$\forall x \in X \ldots$" and "$\exists y \in Y \ldots$" for suitable specified sets $X, Y$. With respect to this, the following remark is particularly relevant.

*Remark 2.24* Before applying *transfer*, all quantifications on subsets "$\forall x \subseteq X \ldots$" or "$\exists x \subseteq X \ldots$" must be reformulated as "$\forall x \in \mathscr{P}(X) \ldots$" and "$\exists x \in \mathscr{P}(X) \ldots$", respectively, where $\mathscr{P}(X) = \{A \mid A \subseteq X\}$ is the *powerset* of $X$. Similarly, all quantifications on functions $f : A \to B$ must be bounded by $\mathrm{Fun}(A, B)$, the *set of functions* from $A$ to $B$. We stress that doing this is crucial because in general ${}^*\mathscr{P}(X) \neq \mathscr{P}({}^*X)$ and ${}^*\mathrm{Fun}(A, B) \neq \mathrm{Fun}({}^*A, {}^*B)$, as we will show in Proposition 2.49.

*Example 2.25* Consider the property: "$<$ is a linear ordering on the set $A$". Notice first that $<$ is a binary relation on $A$, and hence its hyper-extension ${}^*<$ is a binary relation on ${}^*A$. By definition, $<$ is a linear ordering if and only if the following are satisfied:

(a) $\forall x \in A \ (x \not< x)$,
(b) $\forall x, y, z \in A \ (x < y \text{ and } y < z) \Rightarrow x < z$,
(c) $\forall x, y \in A \ (x < y \text{ or } y < x \text{ or } x = y)$.

Notice that the three formulas above are elementary. Then we can apply *transfer* and conclude that: "${}^*<$ is a linear ordering on ${}^*A$."

---

[3]In logic, properties that talks about elements of a given structure are called *first-order properties*; properties about subsets of the given structure are called *second-order*; properties about families of subsets of the given structure are called *third-order*; and so forth.

Whenever confusion is unlikely, some asterisks will be omitted. So, for instance, we will write $+$ to denote both the sum operation on $\mathbb{N}$, $\mathbb{Z}$, $\mathbb{Q}$ and $\mathbb{R}$, and the corresponding operations on $^*\mathbb{N}$, $^*\mathbb{Z}$, $^*\mathbb{Q}$ and $^*\mathbb{R}$, respectively, as given by the hyper-extension $^*+$.

Similarly as in the example above, it is readily verified that the properties of a discretely ordered ring, as well as the properties of a real-closed ordered field, are elementary because they just talk about the *elements* of the given structures. Thus, by a direct application of *transfer*, one obtain the following results, which generalize the properties presented in Sect. 2.1.

**Theorem 2.26**

1. $^*\mathbb{R}$, *endowed with the hyper-extensions of the sum, product, and order on* $\mathbb{R}$, *is a real-closed ordered field.*[4]
2. $^*\mathbb{Z}$ *is an unbounded discretely ordered subring of* $^*\mathbb{R}$, *whose positive part is* $^*\mathbb{N}$.
3. *The ordered subfield* $^*\mathbb{Q} \subset {}^*\mathbb{R}$ *is the quotient field of* $^*\mathbb{Z}$.
4. *Every non-zero* $\nu \in {}^*\mathbb{N}$ *has a successor* $\nu + 1$ *and a predecessor* $\nu - 1$.[5]
5. *For every positive* $\xi \in {}^*\mathbb{R}$ *there exists a unique* $\nu \in {}^*\mathbb{N}$ *with* $\nu \leq \xi < \nu + 1$. *As a result,* $^*\mathbb{N}$ *is unbounded in* $^*\mathbb{R}$.

**Proposition 2.27** $(\mathbb{N}, \leq)$ *is an initial segment of* $(^*\mathbb{N}, \leq)$, *that is, if* $\nu \in {}^*\mathbb{N}\backslash\mathbb{N}$, *then* $\nu > n$ *for all* $n \in \mathbb{N}$,

*Proof* For every $n \in \mathbb{N}$, by *transfer* one obtains the validity of the following elementary formula: "$\forall x \in {}^*\mathbb{N}$ $(x \neq 1$ and $\ldots$ and $x \neq n) \Rightarrow x > n$", and hence the proposition holds.

To get a clearer picture of the situation, examples of *non*-elementary properties that are not preserved under hyper-extensions, are now in order.

*Example 2.28* The property of well-ordering (that is, every nonempty subset has a least element) and of completeness of an ordered set are *not* elementary. Indeed, they concern the *subsets* of the given ordered set. Notice that these properties are not preserved by hyper-extensions. In fact, $\mathbb{N}$ is well-ordered but $^*\mathbb{N}$ is not (e.g., the set of infinite hyper-natural numbers has no least element). The real line $\mathbb{R}$ is complete but $^*\mathbb{R}$ is not (e.g., the set of infinitesimal numbers is bounded with no least upper bound).

*Remark 2.29 Transfer* applies also to the well-ordering property of $\mathbb{N}$, provided one formalizes it as: "*Every nonempty element of* $\mathscr{P}(\mathbb{N})$ *has a least element*". (The property "$X$ has a least element" is elementary: "there exists $x \in X$ such that for every $y \in X$, $x \leq y$.") In this way, one gets: "*Every nonempty element of* $^*\mathscr{P}(\mathbb{N})$ *has a least element*". The crucial point here is that $^*\mathscr{P}(\mathbb{N})$ is not equal

---

[4]Recall that an ordered field is *real closed* if every positive element is a square, and every polynomial of odd degree has a root.

[5]An element $\eta$ is the *successor* of $\xi$ (or $\xi$ is the *predecessor* of $\eta$) if $\xi < \eta$ and there are no elements $\zeta$ with $\xi < \zeta < \eta$.

to $\mathscr{P}(^*\mathbb{N})$ (see Proposition 2.49 below). So, the well-ordering property is *not* an elementary property of $\mathbb{N}$, but it is an elementary property of $\mathscr{P}(\mathbb{N})$. Much the same observations can be made about the completeness property. Indeed, virtually all properties of mathematical objects can be formalized by elementary formulas, provided one uses the appropriate parameters.

A much more slippery example of a *non*-elementary property is the following.

*Example 2.30* The Archimedean property of an ordered field $\mathbb{F}$ is *not* elementary. Notice that to formulate it, one needs to use $\mathbb{N} \subset \mathbb{F}$ as a parameter:

$$\text{"For all positive } x \in \mathbb{F} \text{ there exists } n \in \mathbb{N} \text{ such that } nx > 1.\text{"}$$

While the above is an elementary property of the pair $(\mathbb{F}, \mathbb{N})$ since it talks about the elements of $\mathbb{F}$ and $\mathbb{N}$ combined, it is *not* an elementary property of the ordered field $\mathbb{F}$ alone. In regard to this, we remark that the following expression:

$$\text{"For all positive } x \in \mathbb{F} \text{ it is } x > 1 \text{ or } 2x > 1 \text{ or } 3x > 1 \text{ or } \dots \text{ or } nx > 1 \text{ or } \dots.\text{"}$$

is *not* a formula, because it would consist in an infinitely long string of symbols if written in full. Notice that the Archimedean property is not preserved by hyper-extensions. For instance, $\mathbb{R}$ is Archimedean, but the hyperreal line $^*\mathbb{R}$ is not, being an ordered field that properly extends $\mathbb{R}$ (see Proposition 2.5).

Similarly, the properties of being infinitesimal, finite, or infinite are *not* elementary properties of elements in a given ordered field $\mathbb{F}$, because to formulate them one needs to also consider $\mathbb{N} \subset \mathbb{F}$ as a parameter.

## 2.4   The Ultrapower Model

It is now time to justify what we have seen in the previous sections and show that star maps that satisfy the *transfer* principle do actually exist. Many researchers using nonstandard methods, especially those less familiar with mathematical logic, feel more comfortable in directly working with a model. However we remark that this is not necessary. Rather, it is worth stressing that all one needs in practice is a good understanding of the *transfer* principle and its use, whereas the underlying construction of a specific star map does not play a crucial role.[6] The situation is similar to what happens when working in real analysis: what really matters is that $\mathbb{R}$ is a complete Archimedean ordered field, along with the fact that such a field actually exists; whereas the specific construction of the real line (e.g., by means of

---

[6]There are a few exceptions to this statement, but we will never see them in the combinatorial applications presented in this book.

Dedekind cuts or by a suitable quotient of the set of Cauchy sequences) is irrelevant when developing the theory.

## 2.4.1 The Ultrapower Construction

The ultrapower construction relies on ultrafilters and so, to begin with, let us fix an ultrafilter $\mathcal{U}$ on a set of indices $I$. For simplicity, in the following we will focus on ultrapowers of $\mathbb{R}$. However, the same construction can be carried out by starting with any mathematical structure.

**Definition 2.31** The *ultrapower* of $\mathbb{R}$ modulo the ultrafilter $\mathcal{U}$, denoted $\mathbb{R}^I/\mathcal{U}$, is the quotient of the family of real $I$-sequences $\mathbb{R}^I = \text{Fun}(I, \mathbb{R}) = \{\sigma \mid \sigma : I \to \mathbb{R}\}$ modulo the equivalence relation $\equiv_{\mathcal{U}}$ defined by setting:

$$\sigma \equiv_{\mathcal{U}} \tau \Leftrightarrow \{i \in I \mid \sigma(i) = \tau(i)\} \in \mathcal{U}.$$

Notice that the properties of being a filter on $\mathcal{U}$ guarantee that $\equiv_{\mathcal{U}}$ is actually an equivalence relation. Equivalence classes are denoted by using square brackets: $[\sigma] = \{\tau \in \text{Fun}(I, \mathbb{R}) \mid \tau \equiv_{\mathcal{U}} \sigma\}$. The pointwise sum and product operations on the ring $\text{Fun}(I, \mathbb{R})$ are inherited by the ultrapower. Indeed, it is easily verified that the following operations are well-defined:

$$[\sigma] + [\tau] = [\sigma + \tau] \quad \text{and} \quad [\sigma] \cdot [\tau] = [\sigma \cdot \tau].$$

The order relation $<$ on the ultrapower is defined by putting:

$$[\sigma] < [\tau] \Leftrightarrow \{i \in I \mid \sigma(i) < \tau(i)\} \in \mathcal{U}.$$

**Proposition 2.32** *The ultrapower* $(\mathbb{R}^I/\mathcal{U}, +, \cdot, <, \mathbf{0}, \mathbf{1})$ *is an ordered field.*

*Proof* All properties of an ordered field are directly proved by using the properties of an ultrafilter. For example, to prove that $<$ is a total ordering, one considers the partition $I = I_1 \cup I_2 \cup I_3$ where $I_1 = \{i \in I \mid \sigma(i) < \tau(i)\}$, $I_2 = \{i \in I \mid \sigma(i) = \tau(i)\}$ and $I_3 = \{i \in I \mid \sigma(i) > \tau(i)\}$: exactly one out of the three sets belongs to $\mathcal{U}$, and hence exactly one out of $[\sigma] < [\tau]$, $[\sigma] = [\tau]$, or $[\sigma] > [\tau]$ holds. As another example, let us show that every $[\sigma] \neq \mathbf{0}$ has a multiplicative inverse. By assumption, $A = \{i \in I \mid \sigma(i) = 0\} \notin \mathcal{U}$, and so the complement $A^c = \{i \in I \mid \sigma(i) \neq 0\} \in \mathcal{U}$. Now pick any $I$-sequence $\tau$ such that $\tau(i) = 1/\sigma(i)$ whenever $i \in A^c$. Then $A^c \subseteq \{i \in I \mid \sigma(i) \cdot \tau(i) = 1\} \in \mathcal{U}$, and hence $[\sigma] \cdot [\tau] = \mathbf{1}$. $\qquad\blacksquare$

There is a canonical way of embedding $\mathbb{R}$ into its ultrapower.

**Definition 2.33** The *diagonal embedding* $d : \mathbb{R} \to \mathbb{R}^I/\mathcal{U}$ is the function that associates to every real number $r$ the equivalence class $[c_r]$ of the corresponding constant $I$-sequence $c_r$.

It is readily seen that $d$ is a 1-1 map that preserves sums, products and the order relation. As a result, without loss of generality, we can identify every $r \in \mathbb{R}$ with its diagonal image $d(r) = [c_r]$, and assume that $\mathbb{R} \subseteq \mathbb{R}^I/\mathcal{U}$ is an ordered subfield.

Notice that if $\mathcal{U} = \mathcal{U}_j$ is principal then the corresponding ultrapower $\mathbb{R}^I/\mathcal{U}_j = \mathbb{R}$ is trivial. Indeed, in this case one has $\sigma \equiv_{\mathcal{U}_j} \tau \Leftrightarrow \sigma(j) = \tau(j)$. Thus, every sequence is equivalent to the constant $I$-sequence with value $\sigma(j)$, and the diagonal embedding $d : \mathbb{R} \to \mathbb{R}^I/\mathcal{U}_j$ is onto.[7]

*Remark 2.34* Under the *Continuum Hypothesis*, one can show that for every pair $\mathcal{U}$, $\mathcal{V}$ of non-principal ultrafilters on $\mathbb{N}$, the ordered fields obtained as the corresponding ultrapowers $\mathbb{R}^\mathbb{N}/\mathcal{U}$, $\mathbb{R}^\mathbb{N}/\mathcal{V}$ of $\mathbb{R}$ are isomorphic.[8]

## 2.4.2   Hyper-Extensions in the Ultrapower Model

In this section we will see how the ultrapower $\mathbb{R}^I/\mathcal{U}$ can be made a model of the hyperreal numbers of nonstandard analysis. Let us start by denoting

$$^*\mathbb{R} = \mathbb{R}^I/\mathcal{U}.$$

We now have to show that the ordered field $^*\mathbb{R}$ has all the special features that make it a set of hyperreal numbers. To this end, we will define a *star map* on the family of all sets of ordered tuples of real numbers and of all real functions, in such a way that the *transfer* principle holds.

**Definition 2.35** Let $A \subseteq \mathbb{R}$. Then its *hyper-extension* $^*A \subseteq {}^*\mathbb{R}$ is defined as the family of all equivalence classes of $I$-sequences that take values in $A$, that is:

$$^*A = A^I/\mathcal{U} = \{[\sigma] \mid \sigma : I \to A\} \subseteq {}^*\mathbb{R}.$$

Similarly, if $A \subseteq \mathbb{R}^k$ is a set of real $k$-tuples, then its *hyper-extension* is defined as

$$^*A = \{([\sigma_1], \ldots, [\sigma_k]) \mid (\sigma_1, \ldots, \sigma_k) : I \to A\} \subseteq {}^*\mathbb{R}^k$$

where we denote $(\sigma_1, \ldots, \sigma_k) : i \mapsto (\sigma_1(i), \ldots, \sigma_k(i))$.

---

[7]In most cases, the converse is also true, namely that if the diagonal embedding is onto, then the ultrafilter is principal. A precise statement of the converse would require a discussion of measurable cardinals, taking us too far afield. It suffices to say that when $I$ is an ultrafilter on a countable set, the converse indeed holds.

[8]This is because they are elementarily equivalent $\aleph_1$-saturated structures of cardinality $\aleph_1$ in a finite language, and they have size $\aleph_1$ under the Continuum Hypothesis.

Notice that, by the properties of ultrafilter, for every $\sigma_1, \ldots, \sigma_k, \tau_1, \ldots, \tau_k : I \to \mathbb{R}$, one has

$$\{i \in I \mid (\sigma_1(i), \ldots, \sigma_k(i)) = (\tau_1(i), \ldots, \tau_k(i))\} \in \mathscr{U} \iff \sigma_s \equiv_{\mathscr{U}} \tau_s \text{ for every}$$
$$s = 1, \ldots, k.$$

In consequence, the above equivalence relation is well-defined, and one has that $([\sigma_1], \ldots, [\sigma_n]) \in {}^*A \Leftrightarrow \{i \mid (\sigma_1, \ldots, \sigma_n) \in A\} \in \mathscr{U}$.

We also define the star map on real ordered tuples by setting

$$^*(r_1, \ldots, r_k) = (r_1, \ldots, r_k).$$

Recall that we identified every $r \in \mathbb{R}$ with the equivalence class $[c_r]$ of the corresponding constant sequence and so, by letting $^*r = r = [c_r]$, we have that $A \subseteq {}^*A$ for every $A \subseteq \mathbb{R}^k$.

We have already seen that $^*\mathbb{R}$ is an ordered field that extends the real line. As a result, every rational function $f : \mathbb{R} \to \mathbb{R}$ is naturally extended to a function $^*f : {}^*\mathbb{R} \to {}^*\mathbb{R}$. However, here we are interested in extending *all* real functions $f : A \to B$ where $A$ and $B$ are set of real tuples, to functions $^*f : {}^*A \to {}^*B$. With ultrapowers, this can be done in a natural way.

**Definition 2.36** Let $f : A \to B$ where $A, B \subseteq \mathbb{R}$. Then the *hyper-extension* of $f$ is the function $^*f : {}^*A \to {}^*B$ defined by setting $^*f([\sigma]) = [f \circ \sigma]$ for every $\sigma : I \to A$.

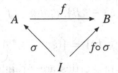

If $f : A \to B$ is a function of several variables where $A \subseteq \mathbb{R}^k$ and $B \subseteq \mathbb{R}$, then $^*f : {}^*A \to {}^*B$ is defined by setting for every $\sigma_1, \ldots, \sigma_k : I \to \mathbb{R}$:

$$^*f([\sigma_1], \ldots, [\sigma_k]) = [\langle f(\sigma_1(i), \ldots, \sigma_k(i)) \mid i \in I \rangle].$$

Similarly as for hyper-extensions of sets of tuples, it is routine to check that the properties of an ultrafilter guarantee that the function $^*f$ is well-defined.

Let us now see that the ultrapower model has all the desired properties.

**Theorem 2.37** *The hyper-extensions of real ordered tuples, sets of ordered real tuples and real functions, as defined above, satisfy all the properties itemized in Propositions 2.15, 2.16, and 2.17. For every $k, n \in \mathbb{N}$, $a, a_1, \ldots, a_k \in \mathbb{R}^n$, and $A, B \subset \mathbb{R}^n$:*

1. *$a = b \Leftrightarrow {}^*a = {}^*b$.*
2. *$a \in A$ if and only if $^*a \in {}^*A$.*

3. *A is a set if and only if* $^*A$ *is a set.*
4. $^*\emptyset = \emptyset$.
5. $A \subseteq B \Leftrightarrow {}^*A \subseteq {}^*B$.
6. $^*(A \cup B) = {}^*A \cup {}^*B$.
7. $^*(A \cap B) = {}^*A \cap {}^*B$.
8. $^*(A \backslash B) = {}^*A \backslash {}^*B$.
9. $^*\{a_1, \ldots, a_k\} = \{a_1, \ldots, a_k\}$.
10. $^*(a_1, \ldots, a_k) = (a_1, \ldots, a_k)$.
11. $^*(A_1 \times \ldots \times A_k) = {}^*A_1 \times \ldots \times {}^*A_k$.
12. $^*\{(a, a) \mid a \in A\} = \{(\xi, \xi) \mid \xi \in A\}$.
13. *R is a k-ary relation if and only if* $^*R$ *is a k-ary relation.*
14. $^*\{a \mid \exists b \, R(a, b)\} = \{\xi \mid \exists \zeta \, ^*R(\xi, \zeta)\}$, *that is,* $^*domain(R) = domain(^*R)$.
15. $^*\{b \mid \exists a \, R(a, b)\} = \{\zeta \mid \exists \xi \, ^*R(\xi, \zeta)\}$, *that is,* $^*range(R) = range(^*R)$.
16. $^*\{(a, b) \mid R(b, a)\} = \{(\xi, \zeta) \mid {}^*R(\zeta, \xi)\}$.
17. $^*\{(a, b, c) \mid S(c, a, b)\} = \{(\xi, \zeta, \eta) \mid {}^*S(\eta, \xi, \zeta)\}$.
18. $^*\{(a, b, c) \mid S(a, c, b)\} = \{(\xi, \zeta, \eta) \mid {}^*S(\xi, \eta, \zeta)\}$.
19. *f is a function if and only if* $^*f$ *is a function.*
20. $^*domain(f) = domain(^*f)$.
21. $^*range(f) = range(^*f)$.
22. $f : A \to B$ *if and only if* $^*f : {}^*A \to {}^*B$.
23. $^*graph(f) = graph(^*f)$.
24. $(^*f)(a) = f(a)$ *for every* $a \in domain(f)$.
25. *If* $f : A \to A$ *is the identity, then* $^*f : {}^*A \to {}^*A$ *is the identity, that is* $^*(1_A) = 1_{^*A}$.
26. $^*\{f(a) \mid a \in A\} = \{^*f(\xi) \mid \xi \in {}^*A\}$, *that is* $^*(f(A)) = {}^*f(^*A)$.
27. $^*\{a \mid f(a) \in B\} = \{\xi \mid {}^*f(\xi) \in {}^*B\}$, *that is* $^*(f^{-1}(B)) = (^*f)^{-1}(^*B)$.
28. $^*(f \circ g) = {}^*f \circ {}^*g$.
29. $^*\{(a, b) \in A \times B \mid f(a) = g(b)\} = \{(\xi, \zeta) \in {}^*A \times {}^*B \mid {}^*f(\xi) = {}^*g(\zeta)\}$.

*Proof* All proofs of the above properties are straightforward applications of the definitions and of the properties of ultrafilters. As an example, let us see here property (13) in detail. We leave the others to the reader as exercises.

Let $\Lambda = \{a \mid \exists b \, R(a, b)\}$ and let $\Gamma = \{\xi \mid \exists \zeta \, ^*R(\xi, \zeta)\}$. We have to show that $^*\Lambda = \Gamma$. If $\sigma : I \to \Lambda$ then for every $i$ there exists an element $\tau(i)$ such that $R(\sigma(i), \tau(i))$. Then $^*R([\sigma], [\tau])$ and so $[\sigma] \in \Gamma$. This shows the inclusion $^*\Lambda \subseteq \Gamma$. Conversely, $[\sigma] \in \Gamma$ if and only if $^*R([\sigma], [\tau])$ for some $I$-sequence $\tau$. Since $([\sigma], [\tau]) \in {}^*R$, the set $\Theta = \{i \mid (\sigma(i), \tau(i)) \in R\} \in \mathscr{U}$, so also the superset $\{i \mid \sigma(i) \in \Lambda\} \supseteq \Theta$ belongs to $\mathscr{U}$. We conclude that $[\sigma] \in {}^*\Lambda$, as desired.

We disclose that the previous theorem essentially states that our defined star map satisfies the *transfer* principle. Indeed, once the notion of elementary property will be made fully rigorous, one can show that *transfer* is actually equivalent to the validity of the properties listed above.

*Remark 2.38* A "strong isomorphism" between two sets of hyperreals $^*\mathbb{R}$ and $^\star\mathbb{R}$ is defined as a bijection $\psi : {}^*\mathbb{R} \to {}^\star\mathbb{R}$ that it coherent with hyper-extensions, that is, $(\xi_1, \ldots, \xi_k) \in {}^*A \Leftrightarrow (\Psi(\xi_1), \ldots, \Psi(\xi_k)) \in {}^*A$ for every $A \subseteq \mathbb{R}^k$ and for every $\xi_1, \ldots, \xi_k \in {}^*\mathbb{R}$, and ${}^*f(\xi_1, \ldots, \xi_k) = \eta \Leftrightarrow {}^*f(\Psi(\xi_1), \ldots, \Psi(\xi_k)) = \Psi(\eta)$ for every $f : \mathbb{R}^k \to \mathbb{R}$ and for every $\xi_1, \ldots, \xi_k, \eta \in {}^*\mathbb{R}$. Then one can show that two ultrapower models $\mathbb{R}^{\mathbb{N}}/\mathscr{U}$ and $\mathbb{R}^{\mathbb{N}}/\mathscr{V}$ are "strongly isomorphic" if and only if the ultrafilters $\mathscr{U} \cong \mathscr{V}$ are isomorphic, that is, there exists a permutation $\sigma : \mathbb{N} \to \mathbb{N}$ such that $A \in \mathscr{U} \Leftrightarrow \sigma(A) \in \mathscr{V}$ for every $A \subseteq \mathbb{N}$. We remark that there exist plenty of non-isomorphic ultrafilters (indeed, one can show that there are $2^c$-many distinct classes of isomorphic ultrafilters on $\mathbb{N}$). This is to be contrasted with the previous Remark 2.34, where the notion of isomorphism between sets of hyperreals was limited to the structure of ordered field.

### 2.4.3 The Properness Condition in the Ultrapower Model

In the previous section, we observed that principal ultrafilters generate trivial ultrapowers. Below, we precisely isolate the class of those ultrafilters that produce models where the properness condition $\mathbb{N} \neq {}^*\mathbb{N}$ (as well as $\mathbb{R} \neq {}^*\mathbb{R}$) holds.

Recall that an ultrafilter $\mathscr{U}$ is called *countably incomplete* if it is not closed under countable intersections, that is, if there exists a countable family $\{I_n\}_{n \in \mathbb{N}} \subseteq \mathscr{U}$ such that $\bigcap_{n \in \mathbb{N}} I_n \notin \mathscr{U}$. We remark that all non-principal ultrafilters on $\mathbb{N}$ or on $\mathbb{R}$ are countably incomplete.[9]

**Exercise 2.39** An ultrafilter $\mathscr{U}$ on $I$ is countably incomplete if and only if there exists a countable partition $I = \bigcup_{n \in \mathbb{N}} J_n$ where $J_n \notin \mathscr{U}$ for every $n$.

**Proposition 2.40** *In the ultrapower model modulo the ultrafilter $\mathscr{U}$ on $I$, the following properties are equivalent:*

1. *Properness condition:* ${}^*\mathbb{N} \neq \mathbb{N}$;
2. *$\mathscr{U}$ is countably incomplete.*

*Proof* Assume first that ${}^*\mathbb{N} \neq \mathbb{N}$. Pick a sequence $\sigma : I \to \mathbb{N}$ such that $[\sigma] \notin \mathbb{N}$. Then $I_n = \{i \in I \mid \sigma(i) \neq n\} \in \mathscr{U}$ for every $n \in \mathbb{N}$, but $\bigcap_n I_n = \emptyset \notin \mathscr{U}$. Conversely, if $\mathscr{U}$ is countably incomplete, pick a countable partition $I = \bigcup_n J_n$ where $J_n \notin \mathscr{U}$ for every $n$, and pick the sequence $\sigma : I \to \mathbb{N}$ where $\sigma(i) = n$ for $i \in J_n$. Then $[\sigma] \in {}^*\mathbb{N}$ but $[\sigma] \neq [c_n]$ for every $n$, where $c_n$ represents the $I$-sequence constantly equal to $n$. ∎

In the sequel we will always assume that ultrapower models are constructed by using ultrafilters $\mathscr{U}$ that are countably incomplete.

---

[9]The existence of non-principal ultrafilters that are countably complete is equivalent to the existence of the so-called *measurable cardinals*, a kind of inaccessible cardinals studied in the hierarchy of large cardinals, and whose existence cannot be proved by ZFC. In consequence, if one sticks to the usual principles of mathematics, it is safe to assume that every non-principal ultrafilter is countably incomplete.

### 2.4.4  An Algebraic Presentation

The ultrapower model can be presented in an alternative, but equivalent, purely algebraic fashion where only the notion of quotient field of a ring modulo a maximal ideal is assumed. Here are the steps of the construction, whose details can be found in [9].

- Consider $\mathrm{Fun}(I, \mathbb{R})$, the ring of real valued sequences where the sum and product operations are defined pointwise.
- Let $\mathfrak{i}$ be the ideal of those sequences that have *finite support*:

$$\mathfrak{i} = \{\sigma \in \mathrm{Fun}(I, \mathbb{R}) \mid \sigma(i) = 0 \text{ for all but at most finitely many } i\}.$$

- Extend $\mathfrak{i}$ to a maximal ideal $\mathfrak{m}$, and define the hyperreal numbers as the quotient field:

$$^*\mathbb{R} = \mathrm{Fun}(I, \mathbb{R})/\mathfrak{m}.$$

- For every subset $A \subseteq \mathbb{R}$, its hyper-extension is defined by:

$$^*A = \{\sigma + \mathfrak{m} \mid \sigma : I \to A\} \subseteq {^*\mathbb{R}}.$$

So, e.g., the *hyper-natural numbers* $^*\mathbb{N}$ are the cosets $\sigma + \mathfrak{m}$ of $I$-sequences $\sigma : I \to \mathbb{N}$ of natural numbers.
- For every function $f : A \to B$ where $A, B \subseteq \mathbb{R}$, its hyper-extension $^*f : {^*A} \to {^*B}$ is defined by setting for every $\sigma : I \to A$:

$$^*f(\sigma + \mathfrak{m}) = (f \circ \sigma) + \mathfrak{m}.$$

It can be directly verified that $^*\mathbb{R}$ is an ordered field whose positive elements are $^*\mathbb{R}^+ = \mathrm{Fun}(\mathbb{N}, \mathbb{R}^+)/\mathfrak{m}$, where $\mathbb{R}^+$ is the set of positive reals. By identifying each $r \in \mathbb{R}$ with the coset $c_r + \mathfrak{m}$ of the corresponding constant sequence, one obtains that $\mathbb{R}$ is a proper subfield of $^*\mathbb{R}$.

Notice that, as in the case of the ultrapower model, the above definitions are naturally extended to hyper-extensions of sets of real tuples and of functions between sets of real tuples.

*Remark 2.41* The algebraic approach presented here is basically equivalent to the ultrapower model. Indeed, for every function $f : I \to \mathbb{R}$, let us denote by $Z(f) = \{i \in I \mid f(i) = 0\}$ its zero-set. If $\mathfrak{m}$ is a maximal ideal of the ring $\mathrm{Fun}(I, \mathbb{R})$, then it is easily shown that the family $\mathscr{U}_{\mathfrak{m}} = \{Z(f) \mid f \in \mathfrak{m}\}$ is an ultrafilter on $\mathbb{N}$. Conversely, if $\mathscr{U}$ is an ultrafilter on $\mathbb{N}$, then $\mathfrak{m}_{\mathscr{U}} = \{f \mid Z(f) \in \mathscr{U}\}$ is a maximal ideal of the ring $\mathrm{Fun}(I, \mathbb{R})$. The correspondence between $\mathscr{U}$-equivalence classes $[\sigma]$ and cosets $\sigma + \mathfrak{m}_{\mathscr{U}}$ yields an isomorphism between the ultrapower $\mathbb{R}^I/\mathscr{U}$ and the quotient $\mathrm{Fun}(I, \mathbb{R})/\mathfrak{m}_{\mathscr{U}}$.

## 2.5  Internal and External Objects

We are now ready to introduce a fundamental class of objects in nonstandard analysis, namely the internal objects. In a way, they are similar to the measurable sets in measure theory, because they are those objects that behave "nicely" in our theory. Indeed, elementary properties of subsets or of functions transfer to the corresponding internal objects (see below).

Recall that the *star map* does not preserve the properties of *powersets* and *function sets*. For instance, we have noticed in the previous sections that there are nonempty) sets in $\mathscr{P}(^*\mathbb{N})$ with no least element, and there are nonempty sets in $\mathscr{P}(^*\mathbb{R})$ that are bounded but have no least upper bound (see Example 2.28 and Remark 2.29). However, by the *transfer principle*, the family $\mathscr{P}(A)$ of all subsets of a set $A$ and $^*\mathscr{P}(A)$ satisfy the same properties. Similarly, the family $\mathrm{Fun}(A, B)$ of all functions $f : A \to B$ and $^*\mathrm{Fun}(A, B)$ satisfy the same properties. Let us now elaborate on this, and start with two easy observations.

**Proposition 2.42**

1. *Every element of the hyper-extension* $^*\mathscr{P}(A)$ *is a subset of* $^*A$, *that is,* $^*\mathscr{P}(A) \subseteq \mathscr{P}(^*A)$;
2. *Every element of the hyper-extension* $^*\mathrm{Fun}(A, B)$ *is a function* $f : {}^*A \to {}^*B$, *that is,* $^*\mathrm{Fun}(A, B) \subseteq \mathrm{Fun}(^*A, {}^*B)$.

*Proof*

(1) Apply *transfer* to the elementary property: $\forall x \in \mathscr{P}(A) \; \forall y \in x \; y \in A$.
(2) Apply *transfer* to the elementary property: $\forall x \in \mathrm{Fun}(A, B)$ "$x$ is a function" and $\mathrm{dom}(x) = A$ and $\mathrm{range}(x) \subseteq B$.

Consequently, it is natural to consider the elements in $^*\mathscr{P}(A)$ as the "nice" subsets of $^*A$, and the elements in $^*\mathrm{Fun}(A, B)$ as the "nice" functions from $^*A$ to $^*B$.

**Definition 2.43** Let $A, B$ be sets. The elements of $^*\mathscr{P}(A)$ are called the *internal subsets* of $^*A$ and the elements of $^*\mathrm{Fun}(A, B)$ are called the *internal functions* from $^*A$ to $^*B$. More generally, an *internal* object is any element $B \in {}^*Y$ that belongs to some hyper-extension.

The following facts about functions are easily verified, and the proofs are left as exercises.

**Proposition 2.44**

1. *A function $F$ is internal if and only if it belongs to the hyper-extension* $^*\mathscr{F}$ *of some set of functions* $\mathscr{F}$;
2. *A function $F : A \to B$ is internal if and only if there exist sets $X, Y$ such that $A \in {}^*\mathscr{P}(X)$, $B \in {}^*\mathscr{P}(Y)$, and $F \in {}^*\{f \text{ function} \mid \mathrm{domain}(f) \subseteq X \text{ and } \mathrm{range}(f) \subseteq Y\}$.*

In consequence, domain and range of an internal function are internal sets.

The first natural examples of internal objects are given by the hyperreal numbers $\xi \in {}^*\mathbb{R}$, and also by all ordered tuples of hyperreal numbers $(\xi_1, \ldots, \xi_k) \in {}^*\mathbb{R}^k$. Notice that the hyper-extension ${}^*X$ of a standard object $X$ is an internal object, since ${}^*X \in {}^*\{X\} = \{{}^*X\}$.

- **Rule of thumb.** Properties about *subsets* of a set $A$ transfer to the *internal subsets* of ${}^*A$, and properties about functions $f : A \rightarrow B$ transfer to the *internal functions* from ${}^*A$ to ${}^*B$.

For instance, the *well-ordering* property of $\mathbb{N}$ is transferred to: "Every nonempty *internal* subset of ${}^*\mathbb{N}$ has a least element", and the *completeness* property of $\mathbb{R}$ transfers to: "Every nonempty *internal* subset of ${}^*\mathbb{R}$ that is bounded above has a least upper bound".

The following is a useful closure property of the class of internal objects.

**Theorem 2.45 (Internal Definition Principle)** *Let $\varphi(x, y_1, \ldots, y_k)$ be an elementary formula. If $A$ is an internal set and $B_1, \ldots, B_n$ are internal objects, then the set $\{x \in A \mid \varphi(x, B_1, \ldots, B_n)\}$ is also internal.*

*Proof* By assumption, there exists a family of sets $\mathscr{F}$ and sets $Y_i$ such that $A \in {}^*\mathscr{F}$ and $B_i \in {}^*Y_i$ for $i = 1, \ldots, n$. Pick any family $\mathscr{G} \supseteq \mathscr{F}$ that is closed under subsets, that is, $C' \subseteq C \in \mathscr{G} \Rightarrow C' \in \mathscr{G}$. (For example, one can take $\mathscr{G} = \bigcup\{\mathscr{P}(C) \mid C \in \mathscr{F}\}$.) Then the following is a true elementary property of the objects $\mathscr{G}, Y_1, \ldots, Y_n$[10]:

$$P(\mathscr{G}, Y_1, \ldots, Y_n): \quad \forall x \in \mathscr{G} \, \forall y_1 \in Y_1 \ldots \forall y_n \in Y_n \, \exists z \in \mathscr{G} \text{ such that}$$

$$\text{``} z = \{t \in x \mid \varphi(t, y_1, \ldots, y_n)\}.\text{''}$$

By *transfer*, the property $P({}^*\mathscr{G}, {}^*Y_1, \ldots, {}^*Y_n)$ is also true, and since $A \in {}^*\mathscr{G}$, $B_i \in {}^*Y_i$, we obtain the existence of an internal set $C \in {}^*\mathscr{G}$ such that $C = \{t \in A \mid \varphi(x, B_1, \ldots, B_n)\}$, as desired. $\square$

As direct applications of the above principle, one obtains the following properties for the class of internal objects.

**Proposition 2.46**

1. *The class $\mathscr{I}$ of internal sets is closed under unions, intersections, set-differences, finite sets and tuples, Cartesian products, and under images and preimages of internal functions.*

---

[10]The subformula "$z = \{t \in x \mid \varphi(t, y_1, \ldots, y_n)\}$" is elementary because it denotes the conjunction of the two formulas:

"$\forall t \in z \, (t \in x \text{ and } \varphi(t, y_1, \ldots, y_n))$" and "$\forall t \in x \, (\varphi(t, y_1, \ldots, y_n) \Rightarrow t \in z)$".

2. *If $A \in \mathcal{I}$ is an internal set, then the set of its internal subsets $\mathcal{P}(A) \cap \mathcal{I} \in \mathcal{I}$ is itself internal.*
3. *If $A$, $B$ are internal sets, then the set $Fun(A, B) \cap \mathcal{I} \in \mathcal{I}$ of internal functions between them is itself internal.*

*Proof*

(1) If $A$ and $B$ are internal sets, say $A \in {}^*\mathcal{P}(X)$ and $B \in {}^*\mathcal{P}(Y)$, then $A \cup B = \{t \in {}^*X \cup {}^*Y \mid t \in A \text{ or } t \in B\}$ is internal by the *Internal Definition Principle*. The other properties are easily proved in the same fashion.
(2) Let $X$ be such that $A \in {}^*\mathcal{P}(X)$. It is easily verified that $\mathcal{P}(A) \cap \mathcal{I} = \{B \in {}^*\mathcal{P}(X) \mid B \subseteq A\}$, and so the *Internal Definition Principle* applies.
(3) Pick $X, Y$ such that $A \in {}^*\mathcal{P}(X)$ and $B \in {}^*\mathcal{P}(Y)$. By Proposition 2.44, we know that

$$Fun(A, B) \cap \mathcal{I} = \{F \in {}^*\mathcal{F} \mid \text{domain}(F) = A \text{ and range}(F) \subseteq B\}$$

where $\mathcal{F} = \{f \text{ function} \mid \text{domain}(f) \subseteq X \text{ and range}(f) \subseteq Y\}$, and so $Fun(A, B) \cap \mathcal{I}$ is internal by the *Internal Definition Principle*.

**Definition 2.47**   An object that is not internal is called *external*.

Although they might not satisfy the same first-order properties as standard sets, external sets are useful in a number of application of nonstandard methods.

*Example 2.48*

1. The set of infinitesimal hyperreal numbers is *external*. Indeed, it is a bounded subset of ${}^*\mathbb{R}$ without least upper bound.
2. The set of infinite hypernatural numbers is *external*. Indeed, it is a nonempty subset of ${}^*\mathbb{N}$ without a least element.
3. The set $\mathbb{N}$ of finite hypernatural numbers is *external*, as its complement ${}^*\mathbb{N} \backslash \mathbb{N}$ inside ${}^*\mathbb{N}$ is external.

The above examples shows that ${}^*\mathcal{P}(\mathbb{N}) \neq \mathcal{P}({}^*\mathbb{N})$ and ${}^*\mathcal{P}(\mathbb{R}) \neq \mathcal{P}({}^*\mathbb{R})$. More generally, we have

**Proposition 2.49**

1. *For every infinite set $A$, the set ${}^\sigma A = \{{}^*a \mid a \in A\}$ is external.*
2. *Every infinite hyperextension ${}^*A$ has external subsets, that is, the inclusion ${}^*\mathcal{P}(A) \subset \mathcal{P}({}^*A)$ is proper.*
3. *If the set $A$ is infinite and $B$ contains at least two elements, then the inclusion ${}^*Fun(A, B) \subset Fun({}^*A, {}^*B)$ is proper.*

*Proof*

(1) Pick a surjective map $\psi : A \to \mathbb{N}$. Then also the hyper-extension ${}^*\psi : {}^*A \to {}^*\mathbb{N}$ is surjective. If by contradiction ${}^\sigma A$ was internal, its image under ${}^*\psi$ would

also be, and this is not possible, since

$$^*\psi\left(^{\sigma}A\right) = \left\{^*\psi(^*a) \mid a \in A\right\} = \left\{^*(\psi(a)) \mid a \in A\right\} = \{\psi(a) \mid a \in A\} = \mathbb{N}.$$

(2) Notice first that $A$ is infinite, because if $A = \{a_1, \ldots, a_n\}$ were finite, then also $^*A = \{^*a_1, \ldots, ^*a_n\}$ would be finite. Recall that $^*\mathscr{P}(A)$ is the set of all internal subsets of $^*A$. Since $^{\sigma}A \subset {}^*A$ is external by (1), $^{\sigma}A \in \mathscr{P}(^*A) \backslash {}^*\mathscr{P}(A)$.

(3) Recall that $^*\mathrm{Fun}(A, B)$ is the set of all internal functions $f : {}^*A \to {}^*B$. Pick an external subset $X \subset A$, pick $b_1 \neq b_2$ in $B$, and let $f : {}^*A \to {}^*B$ be the function where $f(a) = {}^*b_1$ if $a \in X$ and $f(a) = {}^*b_2$ if $a \notin X$. Then $f$ is external, as the preimage $f^{-1}(^*\{b_1\}) = X$ is external.

We warn the reader that becoming familiar with the distinction between internal and external objects is probably the hardest part of learning nonstandard analysis.

### 2.5.1  Internal Objects in the Ultrapower Model

The ultrapower model $^*\mathbb{R} = \mathbb{R}^I / \mathscr{U}$ that we introduced in Sect. 2.4 can be naturally extended so as to include also hyper-extensions of families of sets of real tuples, and of families of functions.

Let us start by observing that every $I$-sequence $T = \langle T_i \mid i \in I \rangle$ of sets of real numbers $T_i \subseteq \mathbb{R}$ determines a set $\widehat{T} \subseteq {}^*\mathbb{R}$ of hyperreal numbers in a natural way, by letting

$$\widehat{T} = \left\{ [\sigma] \in {}^*\mathbb{R} \mid \{i \in I \mid \sigma(i) \in T_i\} \in \mathscr{U} \right\}.$$

**Definition 2.50** If $\mathscr{F} \subseteq \mathscr{P}(\mathbb{R})$, then its *hyper-extension* $^*\mathscr{F} \subseteq {}^*\mathscr{P}(\mathbb{R})$ is defined as

$$^*\mathscr{F} = \left\{ \widehat{T} \mid T : I \to \mathscr{F} \right\}.$$

We remark that the same definition above also applies to families $\mathscr{F} \subseteq \mathscr{P}(\mathbb{R}^k)$ of sets of $k$-tuples, where for $I$-sequences $T : I \to \mathscr{P}(\mathbb{R}^k)$ one defines $\widehat{T} = \left\{ ([\sigma_1], \ldots, [\sigma_k]) \in {}^*\mathbb{R}^k \mid \{i \in I \mid (\sigma_1(i), \ldots, \sigma_k(i)) \in T(i)\} \in \mathscr{U} \right\}$.

According to Definition 2.43, $A \subseteq {}^*\mathbb{R}$ is internal if and only if $A \in {}^*\mathscr{P}(\mathbb{R})$. So, in the ultrapower model, $A \subseteq {}^*\mathbb{R}$ is internal if and only if $A = \widehat{T}$ for some $I$-sequence $T : I \to \mathscr{P}(\mathbb{R})$.

Analogously as above, every $I$-sequence $F = \langle F_i \mid i \in I \rangle$ of real functions $F_i : \mathbb{R} \to \mathbb{R}$ determines a function $\widehat{F} : {}^*\mathbb{R} \to {}^*\mathbb{R}$ on the hyperreal numbers by letting for every $\sigma : I \to \mathbb{R}$:

$$\widehat{F}([\sigma]) = [\langle F_i(\sigma(i)) \mid i \in I \rangle].$$

The internal functions from *$\mathbb{R}$ to *$\mathbb{R}$ in the ultrapower model are precisely those that are determined by some $I$-sequence $F : I \to \mathrm{Fun}(\mathbb{R}, \mathbb{R})$.

**Definition 2.51** If $\mathscr{G} \subseteq \mathrm{Fun}(\mathbb{R}, \mathbb{R})$, then its *hyper-extension* *$\mathscr{G} \subseteq $ *$\mathrm{Fun}(\mathbb{R}, \mathbb{R})$ is defined as

$$*\mathscr{G} = \left\{ \widehat{F} \mid F : I \to \mathscr{G} \right\}.$$

If $F = \langle F_i \mid i \in I \rangle$ is an $I$-sequence of functions $F_i : \mathbb{R}^k \to \mathbb{R}$ of several variables, one extends the above definition by letting $\widehat{F} : $*$\mathbb{R}^k \to $*$\mathbb{R}$ be the function where for every $\sigma_1, \ldots, \sigma_k : I \to \mathbb{R}$:

$$\widehat{F}([\sigma_1], \ldots, [\sigma_k]) = [\langle F_i(\sigma_1(i), \ldots, \sigma_k(i)) \mid i \in I \rangle].$$

Indeed, also in this case, if $\mathscr{G} \subseteq \mathrm{Fun}(\mathbb{R}^k, \mathbb{R})$ then one sets *$\mathscr{G} = \left\{ \widehat{F} \mid F : I \to \mathscr{G} \right\}$.

## 2.6 Hyperfinite Sets

In this section we introduce a fundamental tool in nonstandard analysis, namely the class of *hyperfinite sets*. Although they may contain infinitely many elements, hyperfinite sets satisfy the same "elementary properties" as finite sets. For this reason they are useful in applications as a convenient bridge between finitary statements and infinitary notions.

**Definition 2.52** A *hyperfinite set* $A$ is an element of the hyper-extension *$\mathscr{F}$ of a family $\mathscr{F}$ of finite sets.

In particular, hyperfinite sets are internal objects.

*Remark 2.53* In the ultrapower model, the hyperfinite subsets of *$\mathbb{R}$ are defined according to Definition 2.50. Precisely, $A \subseteq $*$\mathbb{R}$ is hyperfinite if and only if there exists a sequence $\langle T_i \mid i \in I \rangle$ of finite sets $T_i \subset \mathbb{R}$ such that $A = \widehat{T}$, that is, for every $\sigma : I \to \mathbb{R}, [\sigma] \in A \Leftrightarrow \{i \in I \mid \sigma(i) \in T_i\} \in \mathscr{U}$.

Let us start with the simplest properties of hyperfinite sets.

**Proposition 2.54**

1. *A subset $A \subseteq $*$X$ is hyperfinite if and only if $A \in $*$Fin(X)$, where $Fin(X) = \{A \subseteq X \mid A \text{ is finite}\}$.*
2. *Every finite set of internal objects is hyperfinite.*
3. *A set of the form *$X$ for some standard set $X$ is hyperfinite if and only if $X$ is finite.*
4. *If $f : A \to B$ is an internal function, and $\Omega \subseteq A$ is a hyperfinite set, then its image $f(\Omega) = \{$*$f(\xi) \mid \xi \in \Omega\}$ is hyperfinite as well. In particular, internal subsets of hyperfinite sets are hyperfinite.*

*Proof*

(1) If $A$ is a hyperfinite subset of $^*X$, then $A$ is internal, and hence $A \in {}^*\mathscr{P}(X)$. So, if $\mathscr{F}$ is a family of finite sets with $A \in {}^*\mathscr{F}$, then $A \in {}^*\mathscr{P}(X) \cap {}^*\mathscr{F} = {}^*(\mathscr{P}(X) \cap \mathscr{F}) \subseteq {}^*\mathrm{Fin}(X)$. The converse implication is trivial.

(2) Let $A = \{a_1, \ldots, a_k\}$, and pick $X_i$ such that $a_i \in {}^*X_i$. If $X = \bigcup_{i=1}^n X_i$, then $A \in {}^*\mathrm{Fin}(X)$, as it is easily shown by applying *transfer* to the elementary property: "$\forall x_1, \ldots, x_k \in X \{x_1, \ldots, x_k\} \in \mathrm{Fin}(X)$".

(3) This is a direct consequence of *transfer* and the definition of hyperfinite set.

(4) Pick $X$ and $Y$ with $A \in {}^*\mathscr{P}(X)$ and $B \in {}^*\mathscr{P}(Y)$. Then apply *transfer* to the property: "For every $C \in \mathscr{P}(X)$, for every $D \in \mathscr{P}(Y)$, for every $f \in \mathrm{Fun}(C, D)$ and for every $F \in \mathrm{Fin}(X)$ with $F \subseteq C$, the image $f(F)$ is in $\mathrm{Fin}(Y)$".

*Example 2.55* For every pair $N < M$ of (possibly infinite) hypernatural numbers, the interval

$$[N, M]_{*\mathbb{N}} = \{\alpha \in {}^*\mathbb{N} \mid N \leq \alpha \leq M\}$$

is hyperfinite. Indeed, applying *transfer* to the property: "For every $x, y \in \mathbb{N}$ with $x < y$, the set $[x, y]_{\mathbb{N}} = \{a \in \mathbb{N} \mid x \leq a \leq y\} \in \mathrm{Fin}(\mathbb{N})$", one obtains that $[N, M]_{*\mathbb{N}} \in {}^*\mathrm{Fin}(\mathbb{N})$.[11] More generally, it follows from transfer that every bounded internal set of hyperintegers is hyperfinite.

Whenever confusion is unlikely, we will omit the subscript, and write directly $[N, M]$ to denote the interval of hypernatural numbers determined by $N, M \in {}^*\mathbb{N}$.

**Definition 2.56** A *hyperfinite sequence* is an internal function whose domain is a hyperfinite set $A$.

Typical examples of hyperfinite sequences are defined on initial segments $[1, N] \subset {}^*\mathbb{N}$ of the hypernatural numbers. In this case we use notation $\langle \xi_\nu \mid \nu = 1, \ldots, N \rangle$.

By *transfer* from the property: "For every nonempty finite set $A$ there exists a unique $n \in \mathbb{N}$ such that $A$ is in bijection with the segment $\{1, \ldots, n\}$," one obtains that there is a well-posed definition of cardinality for hyperfinite sets.

**Definition 2.57** The *internal cardinality* $|A|_h$ of a nonempty hyperfinite set $A$ is the unique hypernatural number $\alpha$ such that there exists an internal bijection $f : [1, \alpha] \to A$.

**Proposition 2.58** *The internal cardinality satisfies the following properties:*

1. *If the hyperfinite set $A$ is finite, then $|A|_h = |A|$.*
2. *For any $\nu \in {}^*\mathbb{N}$, we have $|[1, \nu]|_h = \nu$. More generally, we have $|[\alpha, \beta]|_h = \beta - \alpha + 1$.*

---

[11]More formally, one transfers the formula: "$\forall x, y \in \mathbb{N} \, [(x < y \Rightarrow (\exists A \in \mathrm{Fin}(\mathbb{N}) \, \forall z \, (z \in A \leftrightarrow x \leq z \leq y))]$".

*Proof*

(1) If $A$ is a finite internal set of cardinality $n$, then every bijection $f : [1, n] \to A$ is internal by Proposition 2.44.
(2) The map $f : [1, \beta - \alpha + 1] \to [\alpha, \beta]$ where $f(i) = \alpha + i - 1$ is an internal bijection.

When confusion is unlikely, we will drop the subscript and directly write $|A|$ to also denote the internal cardinality of a hyperfinite set $A$.

The following is a typical example of a property that hyperfinite sets inherit from finite sets. It is obtained by a straightforward application of *transfer*, and its proof is left as an exercise.

**Proposition 2.59** *Every nonempty hyperfinite subset of* $^*\mathbb{R}$ *has a least element and a greatest element.*

A relevant example of a hyperfinite set which is useful in applications is the following.

**Definition 2.60** Fix an infinite $N \in {}^*\mathbb{N}$. The corresponding *hyperfinite grid* $\mathbb{H}_N \subset$ $^*\mathbb{Q}$ is the hyperfinite set that determines a partition of the interval $[1, N] \subset {}^*\mathbb{R}$ of hyperreals into $N$-many intervals of equal infinitesimal length $1/N$. Precisely:

$$\mathbb{H}_N = \left\{ [1 + \frac{i-1}{N}, 1 + (N-1)\frac{i}{N}] \;\middle|\; i = 1, 2, \ldots, N \right\}.$$

We close this section with a couple of result about the (infinite) cardinalities of hyperfinite sets.

**Proposition 2.61** *If* $\alpha \in {}^*\mathbb{N}$ *is infinite, then the corresponding interval* $[1, \alpha] \subset {}^*\mathbb{N}$ *has cardinality at least the cardinality of the* continuum.

*Proof* For every real number $r \in (0, 1)$, let

$$\psi(r) = \min\{\beta \in [1, \alpha] \mid r < \beta/\alpha\}.$$

Notice that the above definition is well-posed, because $\{\beta \in {}^*\mathbb{N} \mid r < \beta/\alpha\}$ is an internal bounded set of hypernatural numbers, and hence a hyperfinite set. The map $\psi : (0, 1)_{\mathbb{R}} \to [1, \alpha]_{*\mathbb{N}}$ is 1-1. Indeed, $\psi(r) = \psi(s) \Rightarrow |r - s| < 1/\alpha \Rightarrow r \sim s \Rightarrow r = s$ (recall that two real numbers that are infinitely close are necessarily equal). Thus, we obtain the desired inequality $\mathfrak{c} = |(0, 1)_{\mathbb{R}}| \leq |[1, \alpha]_{*\mathbb{N}}|$.

**Corollary 2.62** *If* $A$ *is internal, then either* $A$ *is finite or* $A$ *has at least the cardinality of the* continuum. *In consequence, every countably infinite set is external.*

*Proof* It is easily seen by *transfer* that an internal set $A$ is either hyperfinite, and hence there is an internal bijection with an interval $[1, \alpha] \subset {}^*\mathbb{N}$, or there exists an internal 1-1 function $f : {}^*\mathbb{N} \to A$. In the first case, if $\alpha \in \mathbb{N}$ is finite, then trivially

$A$ is finite. Otherwise $|A| = [1, \alpha] \geq \mathfrak{c}$ by the previous proposition. In the second case, if $\alpha$ is any infinite hypernatural number, then $|A| \geq |^*\mathbb{N}| \geq |[1, \alpha]| \geq \mathfrak{c}$.

### 2.6.1  Hyperfinite Sums

Similarly to finite sums of real numbers, one can consider *hyperfinite* sums of hyperfinite sets of hyperreal numbers.

**Definition 2.63** If $f : A \to \mathbb{R}$ then for every nonempty hyperfinite subset $\Omega \subset {}^*A$, one defines the corresponding *hyperfinite sum* by setting:

$$\sum_{\xi \in \Omega} {}^* f(\xi) := {}^* S_f(\Omega),$$

where $S_f : \mathrm{Fin}(A)\backslash\{\emptyset\} \to \mathbb{R}$ is the function $\{r_1, \ldots, r_k\} \mapsto f(r_1) + \ldots + f(r_k)$.

As a particular case, if $a = \langle a_n \mid n \in \mathbb{N} \rangle$ is a sequence of real numbers and $\alpha \in {}^*\mathbb{N}$ is a hypernatural number, then the corresponding *hyperfinitely long sum* is defined as

$$\sum_{i=1}^{\alpha} a_i = {}^* S_a(\alpha)$$

where $S_a : \mathbb{N} \to \mathbb{R}$ is the function $n \mapsto a_1 + \ldots + a_n$.

*Remark 2.64* More generally, the above definition can be extended to hyperfinite sums $\sum_{\xi \in \Omega} F(\xi)$ where $F : {}^*A \to {}^*\mathbb{R}$ is an internal function, and $\Omega \subseteq {}^*A$ is a nonempty hyperfinite subset. Precisely, in this case one sets $\sum_{\xi \in \Omega} F(\xi) = {}^*\mathscr{S}(F, \Omega)$, where $\mathscr{S} : \mathrm{Fun}(A, \mathbb{R}) \times (\mathrm{Fin}(A)\backslash\{\emptyset\}) \to \mathbb{R}$ is the function $(f, G) \mapsto \sum_{x \in G} f(x)$.

Let us mention in passing that hyperfinite sums can be used to directly define integrals. Indeed, if $N \in {}^*\mathbb{N}$ is any infinite hypernatural number and $\mathbb{H}$ is the corresponding hyperfinite grid (see Definition 2.60), then for every $f : \mathbb{R} \to \mathbb{R}$ and for every $A \subseteq \mathbb{R}$, one defines the *grid integral* by putting:

$$\int_A f(x) d_{\mathbb{H}}(x) = \mathrm{st} \left( \frac{1}{N} \sum_{\xi \in \mathbb{H} \cap {}^*A} {}^* f(\xi) \right).$$

Notice that the above definition applies to *every* real function $f$ and to *every* subset $A$. Moreover, it can be shown that if $f : [a, b] \to \mathbb{R}$ is a Riemann integrable function defined on an interval, then the grid integral coincides with the usual Riemann integral.

## 2.7 Overflow and Underflow Principles

**Proposition 2.65 (Overflow Principles)**

1. $A \subseteq \mathbb{N}$ is infinite if and only if its hyper-extension $^*A$ contains an infinite number.
2. If $B \subseteq {}^*\mathbb{N}$ is internal and $B \cap \mathbb{N}$ is infinite then $B$ contains an infinite number.
3. If $B \subseteq {}^*\mathbb{N}$ is internal and $\mathbb{N} \subseteq B$ then $[1, \alpha] \subseteq B$ for some infinite $\alpha \in {}^*\mathbb{N}$.

*Proof* Item (1) follows from Propositions 2.22 and 2.27. For Item (2), suppose that $B$ does not contain an infinite number. Then $B$ is bounded above in $^*\mathbb{N}$. By *transfer*, $B$ has a maximum, which is necessarily an element of $\mathbb{N}$, contradicting that $B \cap \mathbb{N}$ is infinite. For Item (3), let $C := \{\alpha \in {}^*\mathbb{N} : [1, \alpha] \subseteq B\}$. Then $C$ is internal and $\mathbb{N} \subseteq C$ by assumption. By Item (2) applied to $C$, there is $\alpha \in C$ that is infinite. This $\alpha$ is as desired.

**Proposition 2.66 (Underflow Principles)**

1. If $B \subseteq {}^*\mathbb{N}$ is internal and $B$ contains arbitrarily small infinite numbers, then $B$ contains a finite number.
2. If $B \subseteq {}^*\mathbb{N}$ is internal and $[\alpha, +\infty) \subseteq B$ for every infinite $\alpha \in {}^*\mathbb{N}$ then $[n, +\infty) \subseteq B$ for some finite $n \in \mathbb{N}$.

*Proof* For Item (1), suppose that $B$ does not contain a finite number. Then the minimum of $B$ is necessarily infinite, contradicting the assumption that $B$ contains arbitrarily small infinite numbers. Item (2) follows by applying Item (1) to the internal set $C := \{\alpha \in {}^*\mathbb{N} : [\alpha, +\infty) \subseteq B\}$.

In practice, one often says they are using *overflow* when they are using any of the items in Proposition 2.65 and likewise for *underflow*. Below we will present a use of *overflow* in graph theory.

### 2.7.1 An Application to Graph Theory

Recall that a *graph* is a set $V$ (the set of *vertices*) endowed with an anti-reflexive and symmetric binary relation $E$ (the set of *edges*). Notice that if $G = (V, E)$ is a graph then also its hyper-extension $^*G = (^*V, {}^*E)$ is a graph. By assuming as usual that $^*v = v$ for all $v \in V$, one has that $G$ is a sub-graph of $^*G$. A graph $G = (V, E)$ is *locally finite* if for every vertex $v \in V$, its *set of neighbors* $N_G(v) = \{u \in V \mid \{u, v\} \in E\}$ is finite. One has the following simple nonstandard characterization.

**Proposition 2.67** *A graph $G = (V, E)$ is locally finite if and only if $^*(N_G(v)) \subseteq V$ for every $v \in V$.*

*Proof* If $G$ is locally finite then for every $v \in V$ the set of its neighbors $N_G(v) = \{u_1, \ldots, u_n\}$ is finite, and so $^*N_G(v) = \{^*u_1, \ldots, ^*u_n\} = \{u_1, \ldots, u_n\} \subseteq V$. Conversely, if $G$ is not locally finite, then there exists a vertex $v \in V$ such that $N_G(v)$ is infinite, and we can pick an element $\tau \in {}^*(N_G(v)) \backslash N_G(v)$. Now, $\tau \notin V$, as otherwise $\tau \in {}^*(N_G(v)) \cap V = N_G(v)$, a contradiction.

Recall that a *finite path* in a graph $G = (V, E)$ is a finite sequence $\langle v_i \mid i = 1, \ldots, n \rangle$ of pairwise distinct vertexes such that $\{v_i, v_{i+1}\} \in E$ for every $i < n$. A graph is *connected* if for every pair of distinct vertices $u, u'$ there exists a finite path $\langle v_i \mid i = 1, \ldots, n \rangle$ where $v_1 = u$ and $v_n = u'$. A hyperfinite path $^*G$ is a hyperfinite sequence $\langle v_i \mid i = 1, \ldots, N \rangle$ for $N \in {}^*\mathbb{N}$ of pairwise distinct vertexes $v_i \in {}^*V$ such that $\{v_i, v_{i+1}\} \in {}^*E$ for every $i < n$. An *infinite path* is a sequence $\langle v_i \mid i \in \mathbb{N} \rangle$ of pairwise distinct vertexes such that $\{v_i, v_{i+1}\} \in E$ for every $i \in \mathbb{N}$.

**Theorem 2.68 (König's Lemma - I)** *Every infinite connected graph that is locally finite contains an infinite path.*

*Proof* Given a locally finite connected graph $G = (V, E)$ where $V$ is infinite, pick $u \in V$ and $\tau \in {}^*V \backslash V$. Since $G$ is connected, by *transfer* there exists a hyperfinite sequence $\langle v_i \mid i = 1, \ldots, \mu \rangle$ for some $\mu \in {}^*\mathbb{N}$ where $v_1 = u$ and $\{v_i, v_{i+1}\} \in {}^*E$ for every $i < \mu$. By local finiteness, $^*(N_G(v_1)) \subseteq V$ and so $v_2 \in V$ and $\{v_1, v_2\} \in E$. Then, by induction, it is easily verified that the restriction $\langle v_i \mid i \in \mathbb{N} \rangle$ of the above sequence to the finite indexes is an infinite path in $G$.

A simple but relevant application of *overflow* proves the following equivalent formulation in terms of trees. A graph is a *tree* if between any pair of distinct vertices $u, u'$ there exists a *unique* finite path $\langle v_i \mid i = 1, \ldots, n \rangle$ where $v_1 = u$ and $v_n = u'$. A *rooted tree* is a tree with a distinguished vertex, called the root. The vertices of a tree are also called nodes. The *height* of a node (other than the root) in a rooted tree is the length of the unique finite path that connects is to the root. A *branch* of a rooted tree is an infinite path $\langle v_i \mid i \in \mathbb{N} \rangle$ such that $v_1$ is equal to the root.

**Theorem 2.69 (König's Lemma - II)** *Every infinite, finitely branching tree has an infinite path.*

*Proof* Let $T_n$ denote the nodes of the tree of height $n$. Since $T$ is finitely branching, each $T_n$ is finite. Since $T$ is infinite, each $T_n \neq \emptyset$. By *overflow*, there is $N > \mathbb{N}$ such that $T_N \neq \emptyset$. Fix $x \in T_N$. Then

$\quad \{y \in T \mid$ either $y$ is the root of $T$ or $y$ is connected to $x$ by a hyperfinite path

$\qquad$ that does not pass from the root of $T\}$

is an infinite branch in $T$.

**Exercise 2.70** Prove the last assertion in the proof of Theorem 2.69.

## 2.8 The Saturation Principle

The *transfer principle* is all that one needs to develop the machinery of nonstandard analysis, but for advanced applications another property is also necessary, namely:

**Definition 2.71** *Countable Saturation Principle:* Suppose $\{B_n\}_{n\in\mathbb{N}} \subseteq {}^*A$ is a countable family of internal sets with the finite intersection property. Then $\bigcap_{n\in\mathbb{N}} B_n \neq \emptyset$.

**Exercise 2.72** Assume countable saturation. Then every sequence $\langle B_n \mid n \in \mathbb{N}\rangle$ of internal elements can be extended to an internal sequence $\langle B_n \mid n \in {}^*\mathbb{N}\rangle$, that is, there exists an internal function $\sigma$ with domain ${}^*\mathbb{N}$ and such that $\sigma(n) = B_n$ for every $n \in \mathbb{N}$.

Countable saturation will be instrumental in the definition of *Loeb measures*. In several contexts, stronger saturation principles are assumed where also families of larger size are allowed. Precisely, if $\kappa$ is a given uncountable cardinal, then one considers the following.

**Definition 2.73** $\kappa$-*saturation property*: If $\mathscr{B}$ is a family of internal subsets of ${}^*A$ of cardinality less than $\kappa$, and if $\mathscr{B}$ has the finite intersection property, then $\bigcap_{B\in\mathscr{B}} B \neq \emptyset$.

Notice that, in this terminology, countable saturation is $\aleph_1$-saturation.

In addition to countable saturation, in the applications presented in this book, we will only use the following weakened version of $\kappa$-saturation, where only families of hyper-extensions are considered.

**Definition 2.74** $\kappa$-*enlarging property*: Suppose $\mathscr{F} \subseteq \mathscr{P}(A)$ has cardinality $|\mathscr{F}| < \kappa$. If $\mathscr{F}$ has the finite intersection property, then $\bigcap_{F\in\mathscr{F}} {}^*F \neq \emptyset$.[12]

As a first important application of the enlarging property, one obtains that sets are included in a hyperfinite subset of their hyper-extension.

**Proposition 2.75** *If the* $\kappa$-*enlarging property holds, then for every set $X$ of cardinality $|X| < \kappa$ there exists a hyperfinite subset $H \subseteq {}^*X$ such that $X \subseteq H$.*

*Proof* For each $a \in X$, let $\mathscr{X}_a := \{Y \subseteq X \ : \ Y \text{ is finite and } a \in Y\}$. One then applies the $\kappa$-enlarging property to the family $\mathscr{F} := \{\mathscr{X}_a \ : \ a \in X\}$ to obtain $H \in \bigcap_{a\in X} {}^*\mathscr{X}_a$. Such $H$ is as desired.

The saturation property plays a key role in the application of nonstandard methods to topology, as the next example shows.

---

[12]We remark that the enlarging property is strictly weaker than saturation, in the sense that for every infinite $\kappa$ there are models of nonstandard analysis where the $\kappa$-enlarging property holds but $\kappa$-saturation fails.

*Example 2.76* Let $(X, \tau)$ be a topological space with *character* $< \kappa$, that is, such that each point $x \in X$ has a base of neighborhoods $\mathcal{N}_x$ of cardinality less than $\kappa$. If we assume the $\kappa$-enlarging property, the intersection $\mu(x) = \bigcap_{U \in \mathcal{N}_x} {}^*U$ is nonempty. In the literature, $\mu(x)$ is called the *monad* of $x$. Monads are the basic ingredient in applying nonstandard analysis to topology, starting with the following characterizations (see, e.g., [91, Ch.III]):

- $X$ is *Hausdorff* if and only if distinct points of $x$ have disjoint monads;
- $A \subseteq X$ is *open* if and only if for every $x \in A$, $\mu(a) \subseteq {}^*A$;
- $C \subseteq X$ is *closed* if and only if for every $x \notin C$, $\mu(x) \cap {}^*C = \emptyset$;
- $K \subseteq X$ is *compact* if and only if ${}^*K \subseteq \bigcup_{x \in K} \mu(x)$.

### 2.8.1  Saturation in the Ultrapower Model

We now show that the ultrapower model ${}^*\mathbb{R} = \mathbb{R}^I / \mathcal{U}$ introduced in Sect. 2.4 provides an example of nonstandard map that satisfies saturation. Let us start with a direct combinatorial proof in the case of ultrapowers modulo ultrafilters on $\mathbb{N}$.

**Theorem 2.77** *For every non-principal ultrafilter $\mathcal{U}$ on $\mathbb{N}$, the corresponding ultrapower model satisfies countable saturation.*

*Proof* Let $\{B_n\}$ be a countable family of internal subsets of ${}^*\mathbb{R}$ with the finite intersection property. For every $n$, pick a function $T_n : \mathbb{N} \to \mathscr{P}(\mathbb{R}) \backslash \emptyset$ such that

$$B_n = \widehat{T_n} = \left\{ [\sigma] \in {}^*\mathbb{R} \mid \{i \in \mathbb{N} \mid \sigma(i) \in T_n(i)\} \in \mathcal{U} \right\}.$$

For any fixed $n$, pick an element $\tau(n) \in T_1(n) \cap \cdots \cap T_n(n)$ if that intersection is nonempty. Otherwise, pick an element $\tau(n) \in T_1(n) \cap \cdots \cap T_{n-1}(n)$ if that intersection is nonempty, and so forth until $\tau(n)$ is defined. By the definition of $\tau$, one has the following property:

- If $T_1(n) \cap \cdots \cap T_k(n) \neq \emptyset$ and $n \geq k$ then $\tau(n) \in T_1(n) \cap \ldots \cap T_k(n)$.

Now let $k$ be fixed. By the finite intersection property, $\widehat{T_1} \cap \ldots \cap \widehat{T_k} \neq \emptyset$, so there exists $\sigma : \mathbb{N} \to \mathbb{R}$ such that $\Lambda_j = \{i \in \mathbb{N} \mid \sigma(i) \in T_j(i)\} \in \mathcal{U}$ for every $j = 1, \ldots, k$. In particular, the set of indexes $\Gamma(k) = \{i \in \mathbb{N} \mid T_1(i) \cap \ldots \cap T_k(i) \neq \emptyset\} \in \mathcal{U}$ because it is a superset of $\Lambda_1 \cap \ldots \cap \Lambda_k \in \mathcal{U}$. But then the set $\{i \in \mathbb{N} \mid \tau(i) \in T_1(i) \cap \ldots \cap T_k(i)\} \in \mathcal{U}$ because it is a superset of $\{i \in \Gamma(k) \mid i \geq k\} \in \mathcal{U}$. We conclude that $[\tau] \in \widehat{T_1} \cap \ldots \cap \widehat{T_k}$. As this holds for every $k$, the proof is completed.

The above result can be extended to all ultrapower models where the ultrafilter $\mathcal{U}$ on $I$ is *countably incomplete* (recall that every non-principal ultrafilter on $\mathbb{N}$ is countably incomplete).

**Theorem 2.78** *For every infinite cardinal $\kappa$ there exist ultrafilters $\mathcal{U}$ on the set $I =$ Fin($\kappa$) of finite parts of $\kappa$ such that the corresponding ultrapower model satisfies the $\kappa^+$-enlarging property.*

*Proof* For every $x \in \kappa$, let $\widehat{x} = \{a \in I \mid x \in a\}$. Then trivially the family $\mathcal{X} = \{\widehat{x} \mid x \in \kappa\}$ has the finite intersection property. We claim that every ultrafilter $\mathcal{U}$ that extends $\mathcal{X}$ has the desired property.

Suppose that the family $\mathcal{F} = \{B_x \mid x \in \kappa\} \subseteq \mathcal{P}(A)$ satisfies the finite intersection property. Then we can pick a sequence $\sigma : I \to A$ such that $\sigma(a) \in \bigcap_{x \in a} A_x$ for every $a \in I$. The proof is completed by noticing that $[\sigma] \in {}^*A_x$ for every $x \in \kappa$, since $\{a \in I \mid \sigma(a) \in A_x\} \supseteq \widehat{x} \in \mathcal{U}$.

A stronger result holds, but we will not prove it here because it takes a rather technical proof, and we do not need that result in the applications presented in this book. The proof can be found in [25, §6.1].

**Theorem 2.79** *For every infinite cardinal $\kappa$ there exist ultrafilters $\mathcal{U}$ on $\kappa$ (named $\kappa^+$-good ultrafilters) such that the corresponding ultrapower models satisfy the $\kappa^+$-saturation property.*

## 2.9 Hyperfinite Approximation

As established in Proposition 2.75, in sufficiently saturated structures, hyperfinite sets can be conveniently used as "approximations" of infinite structure. The fact that they behave as finite sets makes them particularly useful objects in applications of nonstandard analysis. In this section we will see a few examples to illustrate this. We assume that the nonstandard extension satisfies the $\kappa$-enlarging property, where $\kappa$ is larger than the cardinality of the objects under consideration.

**Theorem 2.80** *Every infinite set can be linearly ordered.*

*Proof* Let $X$ be an infinite set, and let $\kappa$ be the cardinality of $X$. We can assume that the nonstandard map satisfies the $\kappa^+$-enlargement property. By Proposition 2.75 there exists a hyperfinite $H \subseteq {}^*X$ such that $\{{}^*x \mid x \in X\} \subseteq H$. By *transfer* applied to the corresponding property of finite sets, $H$ can be linearly ordered, whence so can $\{{}^*x \mid x \in X\}$, and hence $X$.

The next theorem is a generalization of the previous one:

**Theorem 2.81** *Every partial order on a set can be extended to a linear order.*

*Proof* We leave it as an easy exercise by induction to show that every partial order on a finite set can be extended to a linear order. Thus, we may precede as in the previous theorem. This time, $H$ is endowed with the partial order it inherits from ${}^*X$, whence, by *transfer*, this partial order can be extended to a linear order. This linear order restricted to $X$ extends the original partial order on $X$.

Recall that a $k$-coloring of a graph $G$ for some $k \in \mathbb{N}$ is a function that assigns to each vertex of $G$ and element of $\{1, \ldots, k\}$ (its color) in such a way that vertices connected by an edge are given different colors. A graph is $k$-colorable if and only if it admits a $k$-coloring.

**Theorem 2.82** *A graph is $k$-colorable if and only if every finite subgraph is $k$-colorable.*

*Proof* Suppose that $G$ is a graph such that every finite subgraph is $k$-colorable. Embed $G$ into a hyperfinite subgraph $H$ of $^*G$. By *transfer*, $H$ can be $k$-colored. The restriction of this $k$-coloring to $G$ is a $k$-coloring of $G$.

The next result plays an important role in the application of ultrafilter and nonstandard methods. Let $S$ be a set. Say that $f : S \to S$ is *fixed-point free* if $f(x) \neq x$ for all $x \in S$.

**Theorem 2.83** *Suppose that $f : S \to S$ is fixed-point free. Then there is a function $c : S \to \{1, 2, 3\}$ (that is, a 3-coloring of $S$) such that $c(f(x)) \neq c(x)$ for all $x \in S$.*

*Proof* In order to use hyperfinite approximation, we first need a finitary version of the theorem:

*Claim* For every finite subset $F \subseteq \mathbb{N}$, there is a 3-coloring $c_F$ of $F$ such that $c(f(n)) \neq c(n)$ whenever $n, f(n) \in F$.

*Proof of Claim* We prove the claim by induction on the cardinality of $F$, the case $|F| = 1$ being trivial since $F$ never contains both $n$ and $f(n)$. Now suppose that $|F| > 1$. Fix $m \in F$ such that $|f^{-1}(m) \cap F| \leq 1$. Such an $m$ clearly exists by the Pigeonhole principle. Let $G := F \backslash \{m\}$. By the induction assumption, there is a 3-coloring $c_G$ of $G$ such that $c(f(n)) \neq c(n)$ whenever $n, f(n) \in G$. One extends $c_G$ to a 3-coloring $c_F$ of $F$ by choosing $c_F(m)$ different from $c_G(f(m))$ (if $f(m) \in G$) and different from $c_G(k)$ if $k \in G$ is such that $f(k) = m$ (if there is such $k$). Since we have three colors to choose from, this is clearly possible. The coloring $c_F$ is as desired.

Now that the claim has been proven, let $H \subseteq {}^*S$ be hyperfinite such that $S \subseteq H$. By *transfer*, there is an internal 3-coloring $c_H$ of $H$ such that $c(f(x)) \neq c(x)$ whenever $x, f(x) \in H$. Since $x \in S$ implies $n, f(n) \in H$, we see that the restriction of $c_H$ to $S$ is a 3-coloring of $S$ as desired.

# Notes and References

Nonstandard analysis was introduced by Robinson in the 1960s [112]. Robinson's original approach was based on model theory. Shortly after, Luxemburg proposed an alternative approach based on the ultrapower construction [99], which helped to further popularize nonstandard methods. Indeed, the ultrapower construction is still one of the most common ways to present nonstandard methods. This is the approach followed in [53], which is an accessible introduction to nonstandard analysis, including a rigorous formulation and a detailed proof of the *transfer principle*. The foundations of nonstandard analysis are also presented in detail in §4.4 of [25]. A survey of several different possible approaches to nonstandard methods is given in [11]. A nice introduction to nonstandard methods for number theorists, including many examples, is presented in [80] (see also [76]). Finally, a full development of nonstandard analysis can be found in several monographs in the existing literature; see e.g. Keisler's classical book [83], or the comprehensive collection of surveys in [3].

# Chapter 3
# Hyperfinite Generators of Ultrafilters

In this chapter, we will show that there is a very tight connection between ultrafilters (introduced in Chap. 1) and nonstandard methods (introduced in Chap. 2). Indeed, any element (or "point") in a nonstandard extension gives rise to an ultrafilter and, conversely, any ultrafilter can be obtained in this way. Such an observation makes precise the assertion that ultrafilter proofs can be seen and formulated as nonstandard arguments. The converse is often true but not always, as two different points in a nonstandard extension can give rise to the same ultrafilter.

Throughout this chapter, we fix an infinite set $S$ and we assume that $^*s = s$ for every $s \in S$, so that $S \subseteq {}^*S$.

## 3.1 Hyperfinite Generators

An important observation is that elements of $^*S$ generate ultrafilters on $S$:

**Exercise 3.1** Suppose that $\alpha \in {}^*S$. Set $\mathscr{U}_\alpha := \{A \subseteq S : \alpha \in {}^*A\}$.

1. $\mathscr{U}_\alpha$ is an ultrafilter on $S$.
2. $\mathscr{U}_\alpha$ is principal if and only if $\alpha \in S$.

We call $\mathscr{U}_\alpha$ the *ultrafilter on S generated by* $\alpha$. Note that in the case that $\alpha \in S$, there is no conflict between the notation $\mathscr{U}_\alpha$ in this chapter and the notation $\mathscr{U}_\alpha$ from Chap. 1.

**Exercise 3.2** For $k \in \mathbb{N}$ and $\alpha \in {}^*\mathbb{N}$, show that $k\mathscr{U}_\alpha = \mathscr{U}_{k\alpha}$.[1]

---

[1] Recall from Definition 1.18 that $A \in k\mathscr{U} \Leftrightarrow A/k = \{n \in \mathbb{N} \mid nk \in A\} \in \mathscr{U}$.

© Springer Nature Switzerland AG 2019
M. Di Nasso et al., *Nonstandard Methods in Ramsey Theory and Combinatorial Number Theory*, Lecture Notes in Mathematics 2239,
https://doi.org/10.1007/978-3-030-17956-4_3

Recall from Exercise 1.12 that, for every function $f : S \rightarrow T$ and for every ultrafilter $\mathscr{U}$ on $S$, the *image ultrafilter* $f(\mathscr{U})$ is the ultrafilter on $T$ defined by setting

$$f(\mathscr{U}) = \{B \subseteq T \mid f^{-1}(B) \in \mathscr{U}\}.$$

**Exercise 3.3** Show that $f(\mathscr{U}_\alpha) = \mathscr{U}_{f(\alpha)}$.

Since there are at most $2^{2^{|S|}}$ ultrafilters on $S$, if the nonstandard extension is $\kappa$-saturated for $\kappa > 2^{2^{|S|}}$, then $|^*S| > 2^{2^{|S|}}$ and we see that there must exist distinct $\alpha, \beta \in {}^*S \backslash S$ such that $\mathscr{U}_\alpha = \mathscr{U}_\beta$ (see Proposition 3.6 and Exercise 3.7 below). This leads to the following notion, which is of central importance in Part II of this book.

**Definition 3.4** Given $\alpha, \beta \in {}^*S$, we say that $\alpha$ and $\beta$ are *u-equivalent*, written $\alpha \sim \beta$, if $\mathscr{U}_\alpha = \mathscr{U}_\beta$.

One can reformulate Definition 3.4 as follows. A $k$-coloring of $S$ for some $k \in \mathbb{N}$ is a function that assigns to each element of $S$ and element of $\{1, \ldots, k\}$ (its *color*). A finite coloring of $S$ is a $k$-coloring for some $k \in \mathbb{N}$. Each coloring $c$ of $S$ induces a coloring $^*c$ of $^*S$, which we will still denote by $c$. Then we have that two elements $\alpha, \beta$ of $^*S$ are $u$-equivalent if and only if $c(\alpha) = c(\beta)$ for every finite coloring $c$ of $S$.

The following exercise establishes some useful properties of the notion of $u$-equivalence.

**Exercise 3.5**

1. If $\alpha, \beta \in {}^*S$, then $\alpha \sim \beta$ if and only if $\alpha = \beta$.
2. Suppose that $f : S \rightarrow S$ and $\alpha \sim \beta$. Then $f(\alpha) \sim f(\beta)$.
3. Suppose that $f : S \rightarrow S$ and $\alpha$ is such that $f(\alpha) \sim \alpha$. Then $f(\alpha) = \alpha$. (Hint: prove the contrapositive using the coloring from Theorem 2.83.)

We have seen that elements of $^*S$ generate ultrafilters on $S$. Under sufficient saturation, the converse holds:

**Proposition 3.6** *Assume that the nonstandard universe has the $(2^{|S|})^+$-enlarging property. Then for every $\mathscr{U} \in \beta S$, there is $\alpha \in {}^*S$ such that $\mathscr{U} = \mathscr{U}_\alpha$.*

*Proof* Fix $\mathscr{U} \in \beta S$. It is clear that $\mathscr{U}$ is a family of subsets of $S$ of cardinality $|\mathscr{U}| \leq 2^{|S|}$ with the finite intersection property, whence, by the $(2^{|S|})^+$-enlarging property, there is $\alpha \in \bigcap_{A \in \mathscr{U}} {}^*A$. Observe now that $\mathscr{U} = \mathscr{U}_\alpha$.

**Exercise 3.7** Assume the $(2^{|S|})^+$-enlarging property. Show that for every non-principal $\mathscr{U} \in \beta S \backslash S$ there exist $|^*S|$-many $\alpha \in {}^*S$ such that $\mathscr{U} = \mathscr{U}_\alpha$.

By the previous proposition, the map $\alpha \mapsto \mathscr{U}_\alpha : {}^*S \rightarrow \beta S$ is surjective. This suggests that we define a topology on $^*S$, called the *u-topology on $^*S$*, by declaring the sets $*A$, for $A \subseteq S$, to be the basic open sets.[2] This topology, while (quasi)compact by the enlarging property, is not Hausdorff. In fact, $\alpha, \beta \in *S$ are *not* separated in the $u$-topology precisely when $\alpha \sim \beta$. Passing to the separation, we get a compact Hausdorff space $*S/\sim$ and the surjection $*S \to \beta S$ defined above descends to a homeomorphism between the quotient space $*S/\sim$ and $\beta S$. So, while $\beta S$ is the "largest" Hausdorff compactification of the discrete space $S$, a (sufficiently saturated) hyper-extension of $S$ is an even larger space, which is still compact (but non-Hausdorff) and has $\beta S$ as a quotient.

## 3.2  The Case of a Semigroup Again

Let us now suppose, once again, that $S$ is the underlying set of a semigroup $(S, \cdot)$. One might guess that, for $\alpha, \beta \in *S$, we have that the equation $\mathcal{U}_{\alpha \cdot \beta} = \mathcal{U}_\alpha \odot \mathcal{U}_\beta$ holds. Unfortunately, this is not the case, even in the case when $S$ is the additive semigroup of positive integers $(\mathbb{N}, +)$:

*Example 3.8* Fix any $\alpha \in *\mathbb{N} \backslash \mathbb{N}$. We show that there is $\beta \in *\mathbb{N}$ such that $\mathcal{U}_\alpha \oplus \mathcal{U}_\beta \neq \mathcal{U}_\beta \oplus \mathcal{U}_\alpha$. For this $\beta$, we must have that either $\mathcal{U}_\alpha \oplus \mathcal{U}_\beta \neq \mathcal{U}_{\alpha + \beta}$ or $\mathcal{U}_\beta \oplus \mathcal{U}_\alpha \neq \mathcal{U}_{\beta + \alpha}$.

Let $A = \bigcup_{n \text{ even}} [n^2, (n+1)^2)$. Take $\nu \in *\mathbb{N}$ such that $\nu^2 \le \alpha < (\nu+1)^2$. Without loss of generality, we may assume that $\nu$ is even. (The argument when $\nu$ is odd is exactly the same.) First suppose that $(\nu+1)^2 - \alpha$ is finite. In this case, we let $\beta := \nu^2$. Note that $\{n \in \mathbb{N} : (A - n) \in \mathcal{U}_\alpha\} = \{n \in \mathbb{N} : n + \alpha \in *A\}$ is finite by assumption, whence not in $\mathcal{U}_\beta$. Consequently, $A \notin \mathcal{U}_\beta \oplus \mathcal{U}_\alpha$. However, since $\alpha - \beta$ is necessarily infinite, we have $\{n \in \mathbb{N} : (A - n) \in \mathcal{U}_\beta\} = \{n \in \mathbb{N} : n + \beta \in *A\} = \mathbb{N}$, whence a member of $\mathcal{U}_\alpha$ and thus $A \in \mathcal{U}_\alpha \oplus \mathcal{U}_\beta$. If $(\nu + 1)^2 - \alpha$ is infinite, then set $\beta := (\nu + 1)^2$. An argument analogous to the argument in the previous paragraph shows that $A \notin \mathcal{U}_\alpha \oplus \mathcal{U}_\beta$ but $A \in \mathcal{U}_\beta \oplus \mathcal{U}_\alpha$.

*Remark 3.9* The previous argument also gives a nonstandard proof of the fact that the center of $(\beta \mathbb{N}, \oplus)$ is precisely the set of principal ultrafilters.

The previous example notwithstanding, there is a connection between $(\beta S, \cdot)$ and the nonstandard extension of the semigroup $(S, \cdot)$. Fix $\alpha, \beta \in *S$. Define $A \cdot \mathcal{U}_\beta^{-1}$ to be the set $\{a \in S : \{b \in S : a \cdot b \in A\} \in \mathcal{U}_\beta\}$. For $a \in S$, we have that $a \in A \cdot \mathcal{U}_\beta^{-1}$ if and only if $a \cdot \beta \in *A$. By transfer, we have that $*(A \cdot \mathcal{U}_\beta^{-1}) = \{\gamma \in *S : \gamma \cdot *\beta \in **A\}$ Hence, we have that

$$A \in \mathcal{U}_\alpha \odot \mathcal{U}_\beta \Leftrightarrow \alpha \in *(A \cdot \mathcal{U}_\beta^{-1}) \Leftrightarrow \alpha \cdot *\beta \in **A.$$

---

[2]This topology is usually named "$S$-topology" in the literature of nonstandard analysis, where the "S" stands for "standard".

Wait! What is $**A$? And what is $*\beta$? Well, our intentional carelessness was intended to motivate the need to be able to take nonstandard extensions of nonstandard extensions, that is, to be able to consider *iterated nonstandard extensions*. Once we give this precise meaning in the next chapter, the above informal calculation will become completely rigorous and we have a precise connection between the operation $\odot$ on $\beta S$ and the operation $\cdot$ on $**S$.

We should also mention that it is possible for the equality $\mathcal{U}_\alpha \odot \mathcal{U}_\beta = \mathcal{U}_{\alpha \cdot \beta}$ to be valid. Indeed, this happens when $\alpha$ and $\beta$ are *independent* in a certain sense; see [37].

## Notes and References

The notion of nonstandard generator of an ultrafilter was initially isolated by Luxemburg in [98]. It was later used by Puritz [107, 108] and by Cherlin and Hirschfeld [26] to study the Rudin-Keisler order among ultrafilters. Model theorists will recognize hyperfinite generators of ultrafilters simply as realizations of the types corresponding to the ultrafilters. A survey on hyperfinite generators of ultrafilters and their properties is presented in [37].

# Chapter 4
# Many Stars: Iterated Nonstandard Extensions

We have seen in Chap. 3 how ultrafilters correspond to points in a nonstandard extension. We will see in this chapter how one can describe operations between ultrafilters, such as the Fubini product, in terms of the corresponding nonstandard points.

In the most common approach to nonstandard methods, one assumes that the star map goes from the usual "standard" universe to a different (larger) "nonstandard" universe. We will see in this chapter that one can dispense of this distinction assume that there is just one universe which is mapped to itself by the star map. This has fruitful consequences, as it allows one to apply the nonstandard map not just one, but any finite number of times. This yields the notion of iterated nonstandard extension, which will be crucial in interpreting the Fubini product and other ultrafilter operations as operations on the corresponding nonstandard points.

## 4.1 The Foundational Perspective

As we saw in the previous chapter, it is useful in applications to consider iterated hyper-extensions of the natural numbers, namely $^*\mathbb{N}$, $^{**}\mathbb{N}$, $^{***}\mathbb{N}$, and so forth. A convenient foundational framework where such iterations make sense can be obtained by considering models of nonstandard analysis where the *standard universe* and the *nonstandard universe* coincide.[1] In other words, one works with a *star map*

$$* : \mathbb{V} \to \mathbb{V}$$

---

[1] A construction of such star maps is given in Sect. A.1.4 of the foundational appendix.

© Springer Nature Switzerland AG 2019
M. Di Nasso et al., *Nonstandard Methods in Ramsey Theory
and Combinatorial Number Theory*, Lecture Notes in Mathematics 2239,
https://doi.org/10.1007/978-3-030-17956-4_4

from a universe into itself. Clearly, in this case every hyper-extension $^*X$ belongs to the universe $\mathbb{V}$, so one can apply the *star map* to it, and obtain the "second level" hyper-extension $^{**}X$, and so forth.

Let us stress that the *transfer principle* in this context must be handled with much care. The crucial point to keep in mind is that in the equivalence

$$P(A_1, \ldots, A_n) \iff P(^*A_1, \ldots, ^*A_n),$$

the considered objects $A_1, \ldots, A_n$ could be themselves iterated hyper-extensions. In this case, one simply has to add one more "star". Let us elaborate on this with a few examples.

*Example 4.1*  Recall that $\mathbb{N}$ is an initial segment of $^*\mathbb{N}$, that is,

$$\mathbb{N} \subset {}^*\mathbb{N} \text{ and } \forall x \in \mathbb{N} \ \forall y \in {}^*\mathbb{N} \backslash \mathbb{N} \ \ x < y.$$

Thus, by *transfer*, we obtain that:

$$^*\mathbb{N} \subset {}^{**}\mathbb{N} \text{ and } \forall x \in {}^*\mathbb{N} \ \forall y \in {}^{**}\mathbb{N} \backslash {}^*\mathbb{N} \ \ x < y.$$

This means that $^*\mathbb{N}$ is a proper initial segment of the double hyper-image $^{**}\mathbb{N}$, that is, every element of $^{**}\mathbb{N} \backslash {}^*\mathbb{N}$ is larger than all element in $^*\mathbb{N}$.

*Example 4.2*  If $\eta \in {}^*\mathbb{N} \backslash \mathbb{N}$, then by *transfer* $^*\eta \in {}^{**}\mathbb{N} \backslash {}^*\mathbb{N}$, and hence $\eta < {}^*\eta$. Then, again by *transfer*, one obtains that the elements $^*\eta, {}^{**}\eta \in {}^{***}\mathbb{N}$ are such that $^*\eta < {}^{**}\eta$, and so forth.

The above example clarifies that the simplifying assumption $^*r = r$ that was adopted for every $r \in \mathbb{R}$ cannot be extended to hold for all hypernatural numbers . Indeed, we just proved that $\eta \neq {}^*\eta$ for every $\eta \in {}^*\mathbb{N} \backslash \mathbb{N}$.

*Example 4.3*  Since $\mathbb{R} \subset {}^*\mathbb{R}$, by *transfer* it follows that $^*\mathbb{R} \subset {}^{**}\mathbb{R}$. If $\varepsilon \in {}^*\mathbb{R}$ is a positive infinitesimal, that is, if $0 < \varepsilon < r$ for every positive $r \in \mathbb{R}$, then by *transfer* we obtain that $0 < {}^*\varepsilon < \xi$ for every positive $\xi \in {}^*\mathbb{R}$. In particular, $^*\varepsilon < \varepsilon$.

Recall that, by Proposition 2.19, for every elementary formula $\varphi(x, y_1, \ldots, y_n)$ and for all objects $B, A_1, \ldots, A_n$, one has that

$$^*\{y \in B \mid P(y, A_1, \ldots, A_n)\} = \{y \in {}^*B \mid P(y, {}^*A_1, \ldots, {}^*A_n)\}. \quad (\dagger)$$

Of course one can apply the above property also when (some of) the parameters are hyper-extensions.

*Remark 4.4*  In nonstandard analysis, a hyper-extension $^*A$ is often called a "standard" set. This terminology comes from the fact that—in the usual approaches—one considers a star map $* : \mathbb{S} \rightarrow \mathbb{V}$ between the "standard universe" $\mathbb{S}$ and a "nonstandard universe" $\mathbb{V}$. Objects $A \in \mathbb{S}$ are named "standard" and, with some ambiguity, also their hyper-extensions $^*A$ are named "standard".[2] Let us stress that the name "standard" would be misleading in our framework, where there is just one single universe, namely the universe of *all* mathematical objects. Those objects of our universe that happen to be in the range of the star map, are called hyper-extensions.

## 4.2  Revisiting Hyperfinite Generators

In this subsection, we let $(S, +)$ denote an infinite semigroup. Now that we have the ability to take iterated nonstandard extensions, we can make our discussion from the end of Sect. 3.2 precise. Recall that, for $\alpha \in {}^*S$, we let $\mathscr{U}_\alpha$ denote the ultrafilter $\{A \subseteq S : \alpha \in {}^*A\}$. Similarly, we can define, for $\alpha \in {}^{**}S$, $\mathscr{U}_\alpha$ to be the ultrafilter $\{A \subseteq S : \alpha \in {}^{**}A\}$.

**Proposition 4.5**  *For $\alpha, \beta \in {}^*S$, we have $\mathscr{U}_\alpha \odot \mathscr{U}_\beta = \mathscr{U}_{\alpha \cdot {}^*\beta}$.*

*Proof*  By equation (†) from the previous section, we have that ${}^*(A \cdot \mathscr{U}_\beta^{-1}) = \{\gamma \in {}^*S : \gamma \cdot {}^*\beta \in {}^{**}A\}$. Hence, for $A \subseteq S$, we have that

$$A \in \mathscr{U}_\alpha \odot \mathscr{U}_\beta \Leftrightarrow \alpha \in {}^*(A \cdot \mathscr{U}_\beta^{-1}) \Leftrightarrow \alpha \cdot {}^*\beta \in {}^{**}A.$$

**Exercise 4.6**  The *tensor product* $\mathscr{U} \otimes \mathscr{V}$ of two ultrafilters on $S$ is the ultrafilter on $S \times S$ defined by:

$$\mathscr{U} \otimes \mathscr{V} = \{C \subseteq S \times S \mid \{s \in S \mid C_s \in \mathscr{V}\} \in \mathscr{U}\},$$

where $C_s = \{t \in S \mid (s, t) \in C\}$ is the vertical $s$-fiber of $C$. If $\alpha, \beta \in {}^*S$, prove that $\mathscr{U}_\alpha \otimes \mathscr{U}_\beta = \mathscr{U}_{(\alpha, {}^*\beta)}$.

We can extend this discussion to elements of higher nonstandard iterates of the universe. Indeed, given $\alpha \in {}^{k*}S$, we can define $\mathscr{U}_\alpha := \{A \subseteq S : \alpha \in {}^{k*}A\}$.

**Exercise 4.7**  For $\alpha \in {}^{k*}S$, prove that $\mathscr{U}_\alpha = \mathscr{U}_{*\alpha}$.

For $\alpha, \beta \in \bigcup_k {}^{k*}S$, we define $\alpha \sim \beta$ if and only if $\mathscr{U}_\alpha = \mathscr{U}_\beta$. Note that $\alpha$ and $\beta$ may live in different levels of the iterated nonstandard extensions.

**Exercise 4.8**  Prove that, for $\alpha_0, \ldots, \alpha_k \in {}^*\mathbb{N}$ and $a_0, \ldots, a_k \in \mathbb{N}$, one has

$$a_0 \mathscr{U}_{\alpha_0} \oplus \cdots \oplus a_k \mathscr{U}_{\alpha_k} = \mathscr{U}_{a_0 \alpha_0 + a_1 {}^*\alpha_1 + \cdots + a_k {}^{k*}\alpha_k}.$$

---

[2]To avoid ambiguity, some authors call the hyper-extensions $^*A \in \mathbb{V}$ "internal-standard".

**Exercise 4.9**

1. Suppose that $\alpha, \alpha', \beta, \beta' \in {}^*\mathbb{N}$ are such that $\alpha \sim \alpha'$ and $\beta \sim \beta'$. Prove that $\alpha + {}^*\beta \sim \alpha' + {}^*\beta'$.
2. Find $\alpha, \alpha', \beta, \beta'$ as above with $\alpha + \beta \not\sim \alpha' + \beta'$.

## 4.3   The Iterated Ultrapower Perspective

The ultrapower model does naturally accommodate iterations of hyper-extensions, although one can be easily puzzled when thinking of iterated hyper-extensions in terms of "iterated ultrapowers". Let us try to clarify this point.

Let us fix an ultrafilter $\mathscr{U}$ on $\mathbb{N}$. Since one can take the ultrapower $\mathbb{N}^{\mathbb{N}}/\mathscr{U}$ of $\mathbb{N}$ to get a nonstandard extension of $\mathbb{N}$, it is natural to take an ultrapower $(\mathbb{N}^{\mathbb{N}}/\mathscr{U})^{\mathbb{N}}/\mathscr{U}$ of $\mathbb{N}^{\mathbb{N}}/\mathscr{U}$ to get a further nonstandard extension. The diagonal embedding $d : \mathbb{N}^{\mathbb{N}}/\mathscr{U} \to (\mathbb{N}^{\mathbb{N}}/\mathscr{U})^{\mathbb{N}}/\mathscr{U}$ is the map where $d(\alpha)$ is the equivalence class in $(\mathbb{N}^{\mathbb{N}}/\mathscr{U})^{\mathbb{N}}/\mathscr{U}$ of the sequence that is constantly $\alpha$. We define ${}^*\alpha$ as $d(\alpha)$, but, unlike the first time when we took an ultrapower and identified $n \in \mathbb{N}$ with $d(n)$, let us refrain from identifying $\alpha$ with ${}^*\alpha$. Indeed, recall that, according to the theory developed in the first section of this chapter, ${}^*\alpha$ is supposed to be infinitely larger than $\alpha$. How do we reconcile this fact with the current construction? Well, unlike the first time we took an ultrapower, a new phenomenon has occurred. Indeed, we now have a second embedding $d_0^{\mathscr{U}} : \mathbb{N}^{\mathbb{N}}/\mathscr{U} \to (\mathbb{N}^{\mathbb{N}}/\mathscr{U})^{\mathbb{N}}/\mathscr{U}$ given by taking the ultrapower of the diagonal embedding $d_0 : \mathbb{N} \to \mathbb{N}^{\mathbb{N}}/\mathscr{U}$.[3] Precisely, if $\alpha = [\sigma] \in \mathbb{N}^{\mathbb{N}}/\mathscr{U}$ where $\sigma : \mathbb{N} \to \mathbb{N}$, then $d_0^{\mathscr{U}}(\alpha) = [([c_{\sigma(1)}], [c_{\sigma(2)}], [c_{\sigma(3)}], \ldots)]$. It is thus through this embedding that we identify $\alpha \in \mathbb{N}^{\mathbb{N}}/\mathscr{U}$ with its image $d_0^{\mathscr{U}}(\alpha) \in (\mathbb{N}^{\mathbb{N}}/\mathscr{U})^{\mathbb{N}}/\mathscr{U}$.

It is now straightforward to see that $\alpha < d(\alpha)$ for all $\alpha \in \mathbb{N}^{\mathbb{N}}/\mathscr{U} \backslash \mathbb{N}$. For example, if $\alpha = [(1, 2, 3, \ldots)] \in \mathbb{N}^{\mathbb{N}}/\mathscr{U}$, then we identify $\alpha$ with $[([c_1], [c_2], [c_3], \ldots)] \in (\mathbb{N}^{\mathbb{N}}/\mathscr{U})^{\mathbb{N}}/\mathscr{U}$. Since $[c_n] < \alpha$ for all $n$, we have that $\alpha < [(\alpha, \alpha, \alpha, \ldots)] = d(\alpha) = {}^*\alpha$.

Also, it is also straightforward to see that defining ${}^{**}f$ as $(f^{\mathscr{U}})^{\mathscr{U}}$ extends ${}^*f = f^{\mathscr{U}}$ for any function $f : \mathbb{N} \to \mathbb{N}$. Indeed, if $\alpha = [\sigma] \in \mathbb{N}^{\mathbb{N}}/\mathscr{U}$, then we have that

$$(f^{\mathscr{U}})^{\mathscr{U}}(\alpha) = (f^{\mathscr{U}})^{\mathscr{U}}(d_0^{\mathscr{U}}(\alpha)) = [(f^{\mathscr{U}}([c_{\sigma(1)}]), f^{\mathscr{U}}([c_{\sigma(2)}]), \ldots)]$$

$$= [([c_{f(\sigma(1))}], [c_{f(\sigma(2))}], \ldots)] = d_0^{\mathscr{U}}([f \circ \sigma]) = [f \circ \sigma] = f^{\mathscr{U}}(\alpha).$$

---

[3] Every map $f : A \to B$ yields a natural map $f^{\mathscr{U}} : A^{\mathbb{N}}/\mathscr{U} \to B^{\mathbb{N}}/\mathscr{U}$ between their ultrapowers, by setting $f^{\mathscr{U}}([\sigma]) = [f \circ \sigma]$ for every $\sigma : \mathbb{N} \to A$.

## 4.4  Revisiting Idempotents

Until further notice, fix a semigroup $(S, \cdot)$. Given the correspondence between ultrafilters on $S$ and elements of $^*S$, it is natural to translate the notion of idempotent ultrafilter to the setting of $^*S$. Suppose that $\alpha \in {}^*S$ is such that $\mathscr{U}_\alpha$ is an idempotent ultrafilter on $S$. We thus have that $\mathscr{U}_\alpha = \mathscr{U}_\alpha \odot \mathscr{U}_\alpha = \mathscr{U}_{\alpha \cdot {}^*\alpha}$. This motivates the following:

**Definition 4.10**  $\alpha \in {}^*S$ is *u-idempotent* if $\alpha \cdot {}^*\alpha \sim \alpha$.

We thus see that $\alpha \in {}^*S$ is $u$-idempotent if and only if $\mathscr{U}_\alpha$ is an idempotent ultrafilter on $S$.

As a first example of the nonstandard perspective of idempotents, we offer the following exercise, which gives a nonstandard proof of [15, Theorem 2.10].

**Exercise 4.11**

1. Suppose that $\alpha \in {}^*\mathbb{N}$ is idempotent. Prove that $2\alpha + {}^{**}\alpha$, $2\alpha + {}^*\alpha + {}^{**}\alpha$, and $2\alpha + 2{}^*\alpha + {}^{**}\alpha$ all generate the same ultrafilter, namely $2\mathscr{U}_\alpha \oplus \mathscr{U}_\alpha$.
2. Suppose that $\mathscr{U} \in \beta\mathbb{N}$ is idempotent and $A \in 2\mathscr{U} \oplus \mathscr{U}$. Prove that $A$ contains a 3-termed arithmetic progression. (Hint: Use part (1) and transfer.)

We now seek an analog of the above fact that nonempty closed subsemigroups of $\beta S$ contain idempotents. Suppose that $T \subseteq \beta S$ is a subsemigroup and that $\alpha, \beta \in {}^*S$ are such that $\mathscr{U}_\alpha, \mathscr{U}_\beta \in T$. Since $\mathscr{U}_{\alpha \cdot {}^*\beta} = \mathscr{U}_\alpha \odot \mathscr{U}_\beta \in T$, we are led to the following definition:

**Definition 4.12**  $T \subseteq {}^*S$ is a *u-subsemigroup* if, for any $\alpha, \beta \in T$, there is $\gamma \in T$ such that $\alpha \cdot {}^*\beta \sim \gamma$.

We thus have the following:

**Corollary 4.13**  *Suppose that $T \subseteq {}^*S$ is a nonempty closed u-subsemigroup. Then $T$ contains a u-idempotent element.*

Now suppose instead that $(S, \cdot)$ is a directed partial semigroup. Note that $^*S$ is naturally a partial semigroup with the nonstandard extension of the partial semigroup operation.

**Definition 4.14**  We say that $\alpha \in {}^*S$ is *cofinite* if $s \cdot \alpha$ is defined for every $s \in S$.

We leave it to the reader to check that $\alpha$ is cofinite if and only if $\mathscr{U}_\alpha$ is a cofinite element of $\beta S$. Consequently, Exercise 1.29 implies that any nonempty closed $u$-subsemigroup of the set of cofinite elements of $^*S$ contains an idempotent element.

**Exercise 4.15**  Without using Exercise 1.29, prove that, for any cofinite $\alpha, \beta \in {}^*S$, there is cofinite $\gamma \in {}^*S$ such that $\alpha \cdot {}^*\beta \sim \gamma$. Compare your proof to the proof that $\mathscr{U} \odot \mathscr{V} \in \gamma S$ whenever $\mathscr{U}, \mathscr{V} \in \gamma S$.

## Notes and References

Iterated hyper-extensions and their use to characterize sums of ultrafilters were introduced in 2010 by Di Nasso in unpublished lecture notes, which were eventually included in [37]. The technique of iterated hyper-extensions and its foundations was then systematically studied by Luperi Baglini in his Ph.D. thesis [93], where several applications in combinatorial number theory were also proved. A use of iterated hyper-extensions for a nonstandard proof of Rado's Theorem can be found in [38]. Further applications to the study of partition regularity of equations are obtained in [39, 40, 94–96] (see Chap. 9 below).

# Chapter 5
# Loeb Measure

In finitary combinatorics, one often encounters counting arguments involving the (normalized) counting measure on the finite set under consideration. The continuous analogue of such a basic tool is the Lebesgue measure on $[0, 1]$, $[0, 1]^n$ or, more generally, some other probability space. Such an analogy can be made precise through the nonstandard perspective. Indeed, one can consider a hyperfinite set, such as the interval $[1, N] = \{1, \ldots, N\}$ in $^*\mathbb{N}$ for some hypernatural number $N$, endowed with its *internal counting measure* (defined on the algebra of its internal subsets). Such an internal object in turns gives rise to an (external) probability measure, called Loeb measure. As it turns out, the Lebesgue measure on $[0, 1]$ can be regarded as a restriction of the Loeb measure to a suitable $\sigma$-algebra. This makes precise the intuition that the Lebesgue measure is a limit "at infinity" of normalized counting measures. In this chapter, we will present the construction of the Loeb measure and some of its fundamental properties.

## 5.1 Premeasures and Measures

In this section, we recall some preliminary information from measure theory. Our presentation borrows somewhat from that of Tao [122].

Fix a set $X$. A nonempty set $\mathscr{A} \subseteq \mathscr{P}(X)$ is an *algebra* if it is closed under unions, intersections, and complements, that is, if $A, B \in \mathscr{A}$, then $A \cup B, A \cap B$, and $X \backslash A$ all belong to $\mathscr{A}$. If $\mathscr{A}$ is an algebra of subsets of $X$, then $\emptyset, X \in \mathscr{A}$. An algebra $\mathscr{A}$ on $X$ is said to be a $\sigma$-*algebra* if it is also closed under countable unions, that is, if $A_1, A_2, \ldots$ all belong to $\mathscr{A}$, then so does $\bigcup_{n=1}^{\infty} A_n$. A $\sigma$-algebra is then automatically closed under countable intersections.

© Springer Nature Switzerland AG 2019
M. Di Nasso et al., *Nonstandard Methods in Ramsey Theory
and Combinatorial Number Theory*, Lecture Notes in Mathematics 2239,
https://doi.org/10.1007/978-3-030-17956-4_5

**Exercise 5.1** Suppose that $X$ is a set and $\mathcal{O} \subseteq \mathcal{P}(X)$ is an arbitrary collection of subsets of $X$. Prove that there is a smallest $\sigma$-algebra $\Omega$ containing $\mathcal{O}$. We call this $\sigma$-algebra the $\sigma$-algebra generated by $\mathcal{O}$ and denote it by $\sigma(\mathcal{O})$.

*Remark 5.2* When trying to prove that every element of $\sigma(\mathcal{O})$ has a certain property, one just needs to show that the set of elements having that property contains $\mathcal{O}$ and is a $\sigma$-algebra.

Suppose that $\mathscr{A}$ is an algebra on $X$. A *pre-measure* on $\mathscr{A}$ is a function $\mu : \mathscr{A} \to [0, +\infty]$ satisfying the following two axioms:

- $\mu(\emptyset) = 0$;
- (Countable Additivity) If $A_1, A_2, \ldots$, all belong to $\mathscr{A}$, are pairwise disjoint, *and* $\bigcup_{n=1}^{\infty} A_n$ belongs to $\mathscr{A}$, then $\mu(\bigcup_{n=1}^{\infty} A_n) = \sum_{n=1}^{\infty} \mu(A_n)$.

If $\mathscr{A}$ is a $\sigma$-algebra, then a pre-measure is called a *measure*. If $\mu$ is a measure on $X$ and $\mu(X) = 1$, then we call $\mu$ a *probability measure on* $X$.

**Exercise 5.3** Fix $n \in \mathbb{N}$ and suppose that $X = \{1, 2, \ldots, n\}$. Let $\mathscr{A} := \mathcal{P}(X)$. Then $\mathscr{A}$ is an algebra of subsets of $X$ that is actually a $\sigma$-algebra for trivial reasons. Define the function $\mu : \mathscr{A} \to [0, 1]$ by $\mu(A) = \frac{|A|}{n}$. Then $\mu$ is a probability measure on $\mathscr{A}$, called the *normalized counting measure*.

**Exercise 5.4** Suppose that $\mu : \mathscr{A} \to [0, +\infty]$ is a pre-measure. Prove that $\mu(A) \leq \mu(B)$ for all $A, B \in \mathscr{A}$ with $A \subseteq B$.

For subsets $A, B$ of $X$, we define the *symmetric difference of $A$ and $B$* to be $A \triangle B := (A \backslash B) \cup (B \backslash A)$.

**Exercise 5.5** Suppose that $\mathscr{A}$ is an algebra and $\mu : \sigma(\mathscr{A}) \to [0, \infty]$ is a measure. Prove that, for every $A \in \sigma(\mathscr{A})$ with $\mu(A) < \infty$ and every $\epsilon \in \mathbb{R}^{>0}$, there is $B \in \mathscr{A}$ such that $\mu(A \triangle B) < \epsilon$.

For our purposes, it will be of vital importance to know that a pre-measure $\mu$ on an algebra $\mathscr{A}$ can be extended to a measure on a $\sigma$-algebra $\mathscr{A}_m$ extending $\mathscr{A}$, a process which is known as *Carathéodory extension*. We briefly outline how this is done. The interested reader can consult any good book on measure theory for all the glorious details; see for instance [122, Section 1.7].

Fix an algebra $\mathscr{A}$ of subsets of $X$ and a pre-measure $\mu$ on $\mathscr{A}$. For arbitrary $A \subseteq X$, we define the *outer measure of $A$* to be

$$\mu^+(A) := \inf\{\sum_{n \in \mathbb{N}} \mu(B_n) \mid A \subseteq \bigcup_{n \in \mathbb{N}} B_n, \text{ each } B_n \in \mathscr{A}\}.$$

Note that $\mu^+(A) = \mu(A)$ for all $A \in \mathscr{A}$. Now although $\mu^+$ is defined on all of $\mathcal{P}(X)$ (which is certainly a $\sigma$-algebra), it need not be a measure. However, there is a canonical $\sigma$-sub-algebra $\mathscr{A}_m$ of $\mathcal{P}(X)$, the so-called *Carathéodory measurable* or $\mu^+$-*measurable subsets of* $X$, on which $\mu^+$ is a measure. These are the sets

$A \subseteq X$ such that

$$\mu^+ (E) = \mu^+ (A \cap E) + \mu^+ (E \backslash A)$$

for every other set $E \subset X$. Let us collect the relevant facts here:

**Fact 5.6** *Let $X$ be a set, $\mathscr{A}$ an algebra of subsets of $X$, and $\mu : \mathscr{A} \to [0, \infty]$ a pre-measure on $\mathscr{A}$ with associated outer measure $\mu^+$ and $\sigma$-algebra of $\mu^+$-measurable sets $\mathscr{A}_m$. Further suppose that $\mu$ is $\sigma$-finite, meaning that we can write $X = \bigcup_{n \in \mathbb{N}} X_n$ with each $X_n \in \mathscr{A}$ and $\mu(X_n) < \infty$.*

1. *$\sigma(\mathscr{A}) \subseteq \mathscr{A}_m$ and $\mu^+|\mathscr{A} = \mu$.*
2. *(Uniqueness) If $\mathscr{A}'$ is another $\sigma$-algebra on $X$ extending $\mathscr{A}$ and $\mu' : \mathscr{A}' \to [0, \infty]$ is a measure on $\mathscr{A}'$ extending $\mu$, then $\mu^+$ and $\mu'$ agree on $\mathscr{A}_m \cap \mathscr{A}'$ (and, in particular, on $\sigma(\mathscr{A})$).*
3. *(Completeness) If $A \subseteq B \subseteq X$ are such that $B \in \mathscr{A}_m$ and $\mu^+(B) = 0$, then $A \in \mathscr{A}_m$ and $\mu^+(A) = 0$.*
4. *(Approximation Results)*

   a. *If $A \in \mathscr{A}_m$, then there is $B \in \sigma(\mathscr{A})$ containing $A$ such that $\mu^+(B \backslash A) = 0$. (So $\mathscr{A}_m$ is the completion of $\sigma(\mathscr{A})$.)*
   b. *If $A \in \mathscr{A}_m$ is such that $\mu^+(A) < \infty$, then for every $\epsilon \in \mathbb{R}^{>0}$, there is $B \in \mathscr{A}$ such that $\mu(A \triangle B) < \epsilon$.*
   c. *Suppose that $A \subseteq X$ is such that, for every $\epsilon \in \mathbb{R}^{>0}$, there is $B \in \mathscr{A}$ such that $\mu(A \triangle B) < \epsilon$. Then $A \in \mathscr{A}_m$.*

*Example 5.7 (Lebesgue Measure)* Suppose that $X = \mathbb{R}$ and $\mathscr{A}$ is the collection of *elementary sets*, namely the finite unions of intervals. Define $\mu : \mathscr{A} \to [0, \infty]$ by declaring $\mu(I) = \text{length}(I)$ and $\mu(I_1 \cup \cdots \cup I_n) = \sum_{i=1}^n \mu(I_j)$ whenever $I_1, \ldots, I_n$ are pairwise disjoint. The above outer-measure procedure yields the $\sigma$-algebra $\mathscr{A}_m$, which is known as the $\sigma$-algebra of *Lebesgue measurable subsets of $\mathbb{R}$* and usually denoted by $\mathfrak{M}$. The measure $\mu^+$ is often denoted by $\lambda$ and is referred to as *Lebesgue measure*. The $\sigma$-algebra $\sigma(\mathscr{A})$ in this case is known as the $\sigma$-algebra of *Borel subsets of $\mathbb{R}$*, usually denoted by $\mathscr{B}$. It can also be seen to be the $\sigma$-algebra generated by the open intervals.

## 5.2 The Definition of Loeb Measure

How do we obtain pre-measures in the nonstandard context? Well, we obtain them by looking at normalized counting measures on hyperfinite sets. Suppose that $X$ is a hyperfinite set. We set $\mathscr{A}$ to be the set of *internal* subsets of $X$. Then $\mathscr{A}$ is an algebra of subsets of $X$ that is not (in general) a $\sigma$-algebra. For example, if $X = [1, N] \subseteq {}^*\mathbb{N}$ for some $N \in {}^*\mathbb{N} \backslash \mathbb{N}$, then for each $n \in \mathbb{N}$, $A_n := \{n\}$ belongs to $\mathscr{A}$, but $\bigcup_n A_n = \mathbb{N}$ does not belong to $\mathscr{A}$ as $\mathbb{N}$ is not internal.

If $A \in \mathscr{A}$, then $A$ is also hyperfinite. We thus define a function $\mu : \mathscr{A} \to [0,1]$ by $\mu(A) := \text{st}\left(\frac{|A|}{|X|}\right)$. We claim that $\mu$ is a pre-measure. It is easily seen to be *finitely additive*, that is, $\mu(A_1 \cup \cdots \cup A_n) = \sum_{i=1}^{n} \mu(A_i)$ whenever $A_1, \ldots, A_n \in \mathscr{A}$ are disjoint. But how do we verify countable additivity? The following simple fact shows that in fact countable additivity follows immediately from finite additivity in this context.

**Proposition 5.8** *If $A_1, A_2, \ldots$ all belong to $\mathscr{A}$ and $\bigcup_{n=1}^{\infty} A_n$ also belongs to $\mathscr{A}$, then there is $k \in \mathbb{N}$ such that $\bigcup_{n=1}^{\infty} A_n = \bigcup_{n=1}^{k} A_n$.*

*Proof* Suppose that the statement of the proposition failed. Set $B := \bigcup_{n=1}^{\infty} A_n$ and, for $k \in \mathbb{N}$, set $B_k := \bigcup_{n=1}^{k} A_n$. By assumption, we have that each $B \setminus B_k$ is a nonempty internal set and the countably family of sets $(B \setminus B_k)$ has the finite intersection property. By countable saturation, it follows that $\bigcap_{k=1}^{\infty} B \setminus B_k = \emptyset$, which is absurd.

We may thus apply the Carathéodory extension theorem from the previous section to obtain a probability measure $\mu^+ : \mathscr{L}_X \to [0,1]$ extending $\mu$. The measure $\mu^+$ is called the *Loeb measure on* $X$ and will be denoted $\mu_X$. The elements of $\mathscr{L}_X$ are referred to as the *Loeb measurable subsets of* $X$.

**Lemma 5.9** *If $B \in \mathscr{L}_X$, then*

$$\mu_X(B) = \inf\{\mu_X(A) \mid A \text{ is internal and } B \subseteq A\}.$$

*Proof* The inequality $\leq$ is clear. Towards the other inequality, fix $\epsilon \in \mathbb{R}^{>0}$. We need to find internal $A$ such that $B \subseteq A$ and $\mu_X(A) \leq \mu_X(B) + \epsilon$. Fix an increasing sequence of internal sets $(A_n \mid n \in \mathbb{N})$ such that $B \subseteq \bigcup_{n \in \mathbb{N}} A_n$ and $\mu_X(A_n) < \mu_X(B) + \epsilon$ for every $n \in \mathbb{N}$. By countable saturation, we extend this sequence to an internal sequence $(A_n \mid n \in {}^*\mathbb{N})$. By transfer, for each $k \in \mathbb{N}$, we have

$$(\forall n \in {}^*\mathbb{N})(n \leq k \to (A_n \subseteq A_k \text{ and } \mu_X(A_n) < \mu_X(B) + \epsilon)).$$

By overflow, there is $K > \mathbb{N}$ such that $\mu_X(A_K) \leq \mu_X(B) + \epsilon$. This concludes the proof.

**Lemma 5.10** *If $B \in \mathscr{L}_X$, then, for every $\epsilon \in \mathbb{R}^{>0}$, there are internal subsets $C, A$ of $X$ such that $C \subseteq B \subseteq A$ and $\mu_X(A \setminus C) < \epsilon$.*

*Proof* Fix $\epsilon > 0$. By Lemma 5.9 applied to $B$, there is an internal set $A$ containing $B$ such that $\mu_X(A) < \mu_X(B) + \frac{\epsilon}{2}$. By Lemma 5.9 applied to $A \setminus B$, there is an internal set $R$ containing $A \setminus B$ such that $\mu_X(R) < \mu_X(A \setminus B) + \frac{\epsilon}{2} < \epsilon$. Set now $C := A \setminus R$ and observe that $C$ is an internal set contained in $B$. Furthermore we have that $\mu_X(A \setminus C) \leq \mu_X(R) < \epsilon$. This concludes the proof.

There are many interesting things to say about Loeb measure. It is crucial for applications of nonstandard analysis to many different areas of mathematics. More information on the Loeb measure can be found in [1, 3]. We will see later in this book that Loeb measure allows one to treat densities on the natural numbers as measures. This makes tools from measure theory and ergodic theory applicable to combinatorial number theory.

## 5.3  Lebesgue Measure via Loeb Measure

The purpose of this section is to see that Lebesgue measure can be constructed using a suitable Loeb measure. The connection between these measures serves as a useful motivation for the results of Chap. 12 on sumsets of sets of positive density. Our presentation borrows somewhat from that of Goldblatt [53].

**Theorem 5.11** *Suppose that* $N > \mathbb{N}$ *and consider the hyperfinite set* $X := \{0, \frac{1}{N}, \frac{2}{N}, \ldots, \frac{N}{N} = 1\}$ *and the function* st $: X \to [0, 1]$. *Define a* $\sigma$-*algebra* $\mathscr{A}$ *on* $[0, 1]$ *by* $A \in \mathscr{A}$ *if and only if* $\mathrm{st}^{-1}(A) \in \mathscr{L}_X$. *For* $A \in \mathscr{A}$, *define* $\nu(A) := \mu_X(\mathrm{st}^{-1}(A))$. *Then* $\mathscr{A}$ *is the algebra of Lebesgue measurable subsets of* $[0, 1]$ *and* $\nu$ *is Lebesgue measure.*

We outline the proof of this theorem in a series of steps. We denote by $\mathscr{B}$ the $\sigma$-algebra of Borel subsets of $[0, 1]$, by $\mathscr{M}$ the $\sigma$-algebra of measurable subsets of $[0, 1]$, and by $\lambda$ the Lebesgue measure on $\mathscr{M}$.

**Exercise 5.12** Prove that $\mathscr{A}$ is a $\sigma$-algebra and $\nu$ is a measure on $\mathscr{A}$.

**Exercise 5.13** Fix $a, b \in [0, 1]$ with $a < b$.

1. Prove that $X \cap (a, b)^* \in \mathscr{L}_X$ and $\mu_X(X \cap (a, b)^*) = b - a$.
2. Prove that $\mathrm{st}^{-1}((a, b)) = \bigcup_{n \in \mathbb{N}}(X \cap (a + \frac{1}{n}, b - \frac{1}{n})^*)$.
3. Prove that $(a, b) \in \mathscr{A}$ and $\nu((a, b)) = b - a$.

We now use the fact that $\lambda$ is the only probability measure on $\mathscr{B}$ satisfying $\lambda(a, b) = b - a$ and that is invariant under translations modulo 1 to conclude that $\mathscr{B} \subseteq \mathscr{A}$ and $\nu|_{\mathscr{B}} = \lambda|_{\mathscr{B}}$.

**Exercise 5.14** Conclude that $\mathscr{M} \subseteq \mathscr{A}$ and $\nu|_{\mathfrak{M}} = \lambda|_{\mathfrak{M}}$. (Hint: Use Fact 5.6.)

**Exercise 5.15** Show that $\mathscr{A} \subseteq \mathscr{M}$. (Hint: if $B \in \mathscr{A}$, then by Lemma 5.10, there are internal $C, D \subseteq X$ such that $C \subseteq \mathrm{st}^{-1}(B) \subseteq D$ and $\mu_X(D \backslash C) < \epsilon$. Set $C' := \mathrm{st}(C)$ and $D' := [0, 1] \backslash \mathrm{st}(X \backslash D)$. Notice that $C'$ is closed and $D'$ is open, whence $C', D' \in \mathscr{B} \subseteq \mathscr{A}$. Prove that $C \subseteq \mathrm{st}^{-1}(C')$ and $\mathrm{st}^{-1}(D') \subseteq D$. Conclude that $B \in \mathscr{M}$.)

## 5.4  Integration

There is a lot to say about the nonstandard theory of integration. We will focus on the Loeb measure $\mu_X$ obtained from a hyperfinite set $X$. In this section, $X$ always denotes a hyperfinite set.

First, if $F : X \to {}^*\mathbb{R}$ is an internal function such that $F(x)$ is finite for $\mu_X$-almost every $x \in X$, we define $\mathrm{st}(F) : X \to \mathbb{R}$ by $\mathrm{st}(F)(x) := \mathrm{st}(F(x))$ whenever $F(x)$ is finite. (Technically speaking, $\mathrm{st}(F)$ is only defined on a set of measure 1, but we will ignore this minor point.) If $f : X \to \mathbb{R}$ is a function and $F : X \to {}^*\mathbb{R}$ is an internal function such that $f(x) = \mathrm{st}(F)(x)$ for $\mu_X$-almost every $x \in X$, we call $F$ a *lift* of $f$. We first characterize which functions have lifts.

**Proposition 5.16** $f : X \to \mathbb{R}$ *has a lift if and only if $f$ is $\mu_X$-measurable.*

*Proof* If $F$ is a lift of $f$, then for any $r \in \mathbb{R}$, we have

$$\mu_X \left( \{x \in X \ : \ f(x) < r\} \triangle \bigcup_{n \in \mathbb{N}} \left\{ x \in X \ : \ F(x) < r - \frac{1}{n} \right\} \right) = 0.$$

Since the latter set is clearly measurable and $\mu_X$ is a complete measure, it follows that $\{x \in X \ : \ f(x) < r\}$ is measurable, whence $f$ is $\mu_X$-measurable.

For the converse, suppose that $f$ is $\mu_X$-measurable and fix a countable open basis $\{V_n\}$ for $\mathbb{R}$. For $n \in \mathbb{N}$, set $U_n := f^{-1}(V_n) \in \mathscr{L}_X$. By Lemma 5.10, one can find, for every $n \in \mathbb{N}$, an increasing sequence $(A_{n,m})$ of internal subsets of $U_n$ such that $\mu_X(A_{n,m}) \geq \mu_X(U_n) - 2^{-m}$ for every $m \in \mathbb{N}$. It follows that the subset

$$X_0 := X \backslash \bigcup_{n \in \mathbb{N}} \left( U_n \backslash \bigcup_{m \in \mathbb{N}} A_{n,m} \right)$$

of $X$ has $\mu_X$-measure 1. Observe now that, for every $n, m \in \mathbb{N}$, there exists an internal function $F : X \to {}^*\mathbb{R}$ such that $F(A_{\ell,k}) \subset {}^*V_\ell$ for $k \leq m$ and $\ell \leq n$. Therefore, by saturation, there exists an internal function $F : X \to {}^*\mathbb{R}$ such that $F(A_{n,m}) \subset {}^*V_n$ for every $n, m \in \mathbb{N}$. It is clear that $f(x) = \mathrm{st}(F(x))$ for every $x \in X_0$, whence $F$ is a lift of $f$. $\qquad\square$

The rest of this section is devoted towards understanding $\int f d\mu_X$ (in the case that $f$ is $\mu_X$-integrable) and the "internal integral" $\frac{1}{|X|} \sum_{x \in X} F(x)$ of a lift $F$ of $f$. We first treat a special, but important, case.

**Lemma 5.17** *Suppose that $F : X \to {}^*\mathbb{R}$ is an internal function such that $F(x)$ is finite for all $x \in X$. Then $\mathrm{st}(F)$ is $\mu_X$-integrable and*

$$\int \mathrm{st}(F) d\mu_X = \mathrm{st} \left( \frac{1}{|X|} \sum_{x \in X} F(x) \right).$$

*Proof* Note first that the assumptions imply that there is $m \in \mathbb{N}$ such that $|F(x)| \leq m$ for all $x \in X$. It follows that $\mathrm{st}(F)$ is $\mu_X$-integrable. Towards establishing the displayed equality, note that, by considering positive and negative parts, that we may assume that $F$ is nonnegative. Fix $n \in \mathbb{N}$. For $k \in \{0, 1, \ldots, mn - 1\}$, set $A_k := \{x \in X : \frac{k}{n} \leq F(x) < \frac{k+1}{n}\}$, an internal set. Since $\sum_k \frac{k}{n} \chi_{A_k}$ is a simple function below $\mathrm{st}(F)$, we have that $\sum_k \frac{k}{n} \mu_X(A_k) \leq \int \mathrm{st}(F) d\mu_X$. However, we also have

$$\sum_k \frac{k}{n} \mu_X(A_k) = \mathrm{st}\left(\frac{1}{|X|} \sum_k \sum_{x \in A_k} \frac{k}{n}\right) \geq \mathrm{st}\left(\frac{1}{|X|} \sum_k \sum_{x \in A_k} (F(x) - \frac{1}{n})\right)$$

$$= \mathrm{st}\left(\frac{1}{|X|} \sum_{x \in X} F(x)\right) - \frac{1}{n}.$$

It follows that $\mathrm{st}(\frac{1}{|X|} \sum_{x \in X} F(x)) \leq \int \mathrm{st}(F) d\mu_X + \frac{1}{n}$. Since $n$ was arbitrary, we have that $\mathrm{st}(\frac{1}{|X|} \sum_{x \in X} F(x)) \leq \int \mathrm{st}(F) d\mu_X$.

The inequality $\int \mathrm{st}(F) d\mu_X \leq \mathrm{st}(\frac{1}{|X|} \sum_{x \in X} F(x))$ is proved in a similar fashion, by considering the simple function $\sum_k \frac{k+1}{n} \chi_{A_k}$.

We now seek to extend the previous lemma to cover situations when $F$ is not necessarily bounded by a standard number. Towards this end, we need to introduce the appropriate nonstandard integrability assumption. A $\mu_X$-measurable internal function $F : X \to {}^*\mathbb{R}$ is called *S-integrable* if:

1. The quantity

$$\frac{1}{|X|} \sum_{x \in X} |F(x)|$$

   is finite, and
2. for every internal subset $A$ of $X$ with $\mu_X(A) = 0$, we have

$$\frac{1}{|X|} \sum_{x \in A} |F(x)| \approx 0.$$

Here is the main result of this section:

**Theorem 5.18** *Suppose that $f : X \to \mathbb{R}$ is a $\mu_X$-measurable function. Then $f$ is $\mu_X$-integrable if and only if $f$ has an S-integrable lifting. In this case, for any S-integrable lift $F$ of $f$ and any internal subset $B$ of $X$, we have*

$$\int_B f d\mu_X = \mathrm{st}\left(\frac{1}{|X|} \sum_{x \in B} F(x)\right).$$

*Proof* We first note that, by taking positive and negative parts, we may assume that $f$ is nonnegative. Moreover, by replacing $f$ with $f \cdot \chi_B$, we may assume that $B = X$.

We first suppose that $F : X \to {}^*\mathbb{R}$ is a nonnegative $S$-integrable function such that $F(x)$ is finite for $\mu_X$-almost every $x$. For $n \in {}^*\mathbb{N}$, set $B_n := \{x \in X : F(x) \geq n\}$.

*Claim 1* For every infinite $N \in {}^*\mathbb{N}$, we have

$$\frac{1}{|X|} \sum_{x \in B_N} F(x) \approx 0.$$

*Proof of Claim 1* Observe that

$$\frac{N |B_N|}{|X|} \leq \frac{1}{|X|} \sum_{x \in B_N} F(x) \leq \frac{1}{|X|} \sum_{x \in X} F(x)$$

Therefore

$$\frac{|B_N|}{|X|} \leq \frac{1}{N} \frac{1}{|X|} \sum_{x \in X} F(x) \approx 0$$

since, by assumption, $\frac{1}{|X|} \sum_{x \in X} F(x)$ is finite. It follows from the assumption that $F$ is $S$-integrable that

$$\frac{1}{|X|} \sum_{x \in B_N} F(x) \approx 0.$$

In the rest of the proof, we will use the following notation: given a nonegative internal function $F : X \to {}^*\mathbb{R}$ and $m \in {}^*\mathbb{N}$, we define the internal function $F_m : X \to {}^*\mathbb{R}$ by $F_m(x) = \min\{F(x), m\}$. Observe that $F_m(x) \leq F_{m+1}(x) \leq F(x)$ for every $m \in {}^*\mathbb{N}$ and every $x \in X$. It follows from the Monotone Convergence Theorem and the fact that, for $\mu_X$-almost every $x \in X$, the sequence $(\mathrm{st}(F_m(x)) : m \in \mathbb{N})$ converges to $\mathrm{st}(F(x))$, that $\int \mathrm{st}(F_m) d\mu_X \to \int \mathrm{st}(F) d\mu_X$.

*Claim 2* We have

$$\mathrm{st}\left(\frac{1}{|X|} \sum_{x \in X} F(x)\right) = \lim_{m \to +\infty} \mathrm{st}\left(\frac{1}{|X|} \sum_{x \in X} F_m(x)\right).$$

*Proof of Claim 2* It is clear that

$$\lim_{m \to \infty} \mathrm{st}\left(\frac{1}{|X|} \sum_{x \in X} F_m(x)\right) \leq \mathrm{st}\left(\frac{1}{|X|} \sum_{x \in X} F(x)\right).$$

For the other inequality, fix $M \in {}^*\mathbb{N}$ infinite and observe that

$$\frac{1}{|X|} \sum_{x \in X} F(x) = \frac{1}{|X|} \sum_{x \in B_M} F(x) + \frac{1}{|X|} \sum_{x \in X \setminus B_M} F(x)$$

$$\approx \frac{1}{|X|} \sum_{x \in X \setminus B_M} F(x)$$

$$= \frac{1}{|X|} \sum_{x \in X \setminus B_M} F_M(x)$$

$$\leq \frac{1}{|X|} \sum_{x \in X} F_M(x).$$

Thus, given any $\epsilon > 0$, we have that $\frac{1}{|X|} \sum_{x \in X} F(x) \leq \frac{1}{|X|} \sum_{x \in X} F_M(x) + \epsilon$ for all infinite $M$, whence, by underflow, we have that $\frac{1}{|X|} \sum_{x \in X} F(x) \leq \frac{1}{|X|} \sum_{x \in X} F_m(x) + \epsilon$ for all but finitely many $m \in \mathbb{N}$. It follows that $\mathrm{st}\left(\frac{1}{|X|} \sum_{x \in X} F(x)\right) \leq \lim_{m \to +\infty} \mathrm{st}\left(\frac{1}{|X|} \sum_{x \in X} F_m(x)\right)$, as desired.

By Lemma 5.17, Claim 2, and the discussion preceding Claim 2, we have that $\mathrm{st}(F)$ is $\mu_X$-integrable and $\int \mathrm{st}(F) d\mu = \mathrm{st}\left(\frac{1}{|X|} \sum_{x \in X} F(x)\right)$, as desired.

We now suppose that $f$ is a nonnegative $\mu_X$ integrable function. We must show that $f$ has an $S$-integrable lifting. Let $F$ be any nonnegative lifting of $f$. Note that, for every infinite $M \in {}^*\mathbb{N}$, that $F_M$ is also a lifting of $f$. We will find an infinite $M \in {}^*\mathbb{N}$ such that $F_M$ is also $S$-integrable.

By the Monotone Convergence Theorem, for every $\epsilon > 0$, we have that

$$\left| \int \mathrm{st}(F) d\mu_X - \int \mathrm{st}(F_m) d\mu_X \right| < \epsilon$$

holds for all but finitely many $m \in \mathbb{N}$. Therefore, by Lemma 5.17, we have that

$$\left| \int \mathrm{st}(F) d\mu_X - \frac{1}{|X|} \sum_{x \in X} F_m(x) \right| < \epsilon$$

holds for all but finitely many $m \in \mathbb{N}$. By transfer, there exists infinite $M \in {}^*\mathbb{N}$ such that

$$\int \mathrm{st}(F) d\mu_X = \mathrm{st}\left(\frac{1}{|X|} \sum_{x \in X} F_M(x)\right)$$

and

$$\int f d\mu_X = \text{st}\left(\frac{1}{|X|}\sum_{x \in X} F_M(x)\right).$$

We show that the function $F_M$ is $S$-integrable. Suppose that $B$ is an internal subset of $X$ such that $\mu_X(B) = 0$. Set

$$r := \text{st}\left(\frac{1}{|X|}\sum_{x \in B} |F_M(x)|\right).$$

We wish to show that $r = 0$. Towards this end, fix $m \in \mathbb{N}$. Then we have that

$$r + \int \text{st}(F_m) d\mu_X = r + \int_{X \backslash B} \text{st}(F_m) d\mu_X \approx r + \frac{1}{|X|}\sum_{x \in X \backslash B} F_m(x)$$

$$\leq r + \frac{1}{|X|}\sum_{x \in X \backslash B} F_M(x) \approx \frac{1}{|X|}\sum_{x \in X} F_M(x) \approx \int \text{st}(F) d\mu_X.$$

Letting $m \to +\infty$, we obtain that $r = 0$, as desired.

**Corollary 5.19** *Suppose $f \in L^1(X, \mathscr{L}_X, \mu_X)$ and $\epsilon > 0$. Then there exists internal functions $F, G : X \to {}^*\mathbb{R}$ such that $F \leq f \leq G$ $\mu_X$-almost everywhere and*

$$\max\left\{\left|\int_B f d\mu_X - \frac{1}{|X|}\sum_{x \in B} F(x)\right|, \left|\int_B f d\mu_X - \frac{1}{|X|}\sum_{x \in B} G(x)\right|\right\} \leq \epsilon$$

*for every internal subset $B$ of $X$.*

*Proof* Let $H : X \to {}^*\mathbb{R}$ be a lifting of $f$. Set $F := H - \epsilon/2$ and $G := H + \epsilon/2$. Since $\text{st}(H(x)) = f(x)$ for $\mu_X$-almost every $x \in X$, we conclude that $F(x) \leq f(x) \leq G(x)$ for $\mu_X$-almost every $x \in X$. Furthermore, if $B$ is an internal subset of $X$, then by Lemma 5.18, we have that

$$\left|\int_B f d\mu_X - \frac{1}{|X|}\sum_{x \in B} F(x)\right| \leq \epsilon/2 + \left|\int_B f d\mu_X - \frac{1}{|X|}\sum_{x \in B} H(x)\right| \leq \epsilon$$

and

$$\left|\int_B f d\mu_X - \frac{1}{|X|}\sum_{x \in B} G(x)\right| \leq \epsilon/2 + \left|\int_B f d\mu_X - \frac{1}{|X|}\sum_{x \in B} H(x)\right| \leq \epsilon.$$

This concludes the proof.

## 5.5   Product Measure

Suppose that $(X, \mathscr{A}_X, \nu_X)$ and $(Y, \mathscr{A}_Y, \nu_Y)$ are two probability measure spaces. We can then form their *product* as follows: first, set $\mathscr{A}$ to be the set of finite unions of rectangles of the form $A \times B$, where $A \in \mathscr{A}_X$ and $B \in \mathscr{A}_Y$. The elements of $\mathscr{A}$ are called *elementary sets*. It is an exercise to show that $\mathscr{A}$ is an algebra of subsets of $X \times Y$ and that every element of $\mathscr{A}$ can be written as a finite union of *disjoint* such rectangles. We can then define a pre-measure $\nu$ on $\mathscr{A}$ by $\mu(\bigcup_{i=1}^{n}(A_i \times B_i)) :=$ $\sum_{i=1}^{n}(\nu_X(A_i) \cdot \nu_Y(B_i))$. Applying the outer measure procedure, we get a measure $\nu_X \otimes \nu_Y : \mathscr{A}_m \to [0, 1]$ extending $\nu$. We denote $\mathscr{A}_m$ by $\mathscr{A}_X \otimes \mathscr{A}_Y$.

The following situation will come up in Chap. 16: suppose that $X$ and $Y$ are hyperfinite sets and we construct the Loeb measure spaces $(X, \mathscr{L}_X, \mu_X)$ and $(Y, \mathscr{L}_Y, \mu_Y)$. We are thus entitled to consider the product measure space $(X \times Y, \mathscr{L}_X \otimes \mathscr{L}_Y, \mu_X \otimes \mu_Y)$. However, $X \times Y$ is itself a hyperfinite set, whence we can consider its Loeb measure space $(X \times Y, \mathscr{L}_{X \times Y, L}, \mu_{X \times Y})$. There is a connection:

**Exercise 5.20** Show that $\mathscr{L}_X \otimes \mathscr{L}_Y$ is a sub-$\sigma$-algebra of $\mathscr{L}_{X \times Y}$ and that $\mu_{X \times Y}|_{(\mathscr{L}_X \otimes \mathscr{L}_Y)} = \mu_X \otimes \mu_Y$.

In the proof of the triangle removal lemma in Chap. 16, we will need to use the following Fubini-type theorem for Loeb measure on a hyperfinite set.

**Theorem 5.21** *Suppose that $X$ and $Y$ are hyperfinite sets and $f : X \times Y \to \mathbb{R}$ is a bounded $\mathscr{L}_{X \times Y}$-measurable function. For $x \in X$, let $f_x : Y \to \mathbb{R}$ be defined by $f_x(y) := f(x, y)$. Similarly, for $y \in Y$, let $f^y : X \to \mathbb{R}$ be defined by $f^y(x) := f(x, y)$. Then:*

1. *$f_x$ is $\mathscr{L}_Y$-measurable for $\mu_X$-almost every $x \in X$;*
2. *$f^y$ is $\mathscr{L}_X$-measurable for $\mu_Y$-almost every $y \in Y$;*
3. *The double integral can be computed as an iterated integral:*

$$\int_{X \times Y} f(x, y) d\mu_{X \times Y}(x, y) = \int_X \left( \int_Y f_x(y) d\mu_Y(y) \right) d\mu_X(x)$$

$$= \int_Y \left( \int_X f^y(x) d\mu_X(x) \right) d\mu_Y(y).$$

*Proof* After taking positive and negative parts, it suffices to consider the case that $f$ is positive. Furthermore, by the Monotone Convergence Theorem, it suffices to consider the case that $f$ is a step function. Then, by linearity, one can restrict to the case that $f = \chi_E$ is the characteristic function of a Loeb measurable set $E \subseteq X \times Y$. Now Lemma 5.10 and a further application of the Monotone Convergence Theorem allows one to restrict to the case that $E$ is internal. In this case, for $x \in X$ we have that $\int_Y \chi_E(x, y) d\mu_Y(y) = \text{st}\left(\frac{|E_x|}{|Y|}\right)$, where $E_x := \{y \in Y : (x, y) \in E\}$. By

Theorem 5.18, we thus have

$$\int_X \left( \int_Y \chi_E(x, y) d\mu_Y(y) \right) d\mu_X(x) \approx \frac{1}{|X|} \sum_{x \in X} \frac{|E_x|}{|Y|} = \frac{|E|}{|X||Y|}$$

$$\approx \int_{X \times Y} \chi_E(x, y) d\mu_{X \times Y}(x, y).$$

The other equality is proved in the exact same way.

## 5.6  Ergodic Theory of Hypercycle Systems

**Definition 5.22** If $(X, \mathscr{B}, \mu)$ is a probability space, we say that a function $T$ : $X \to X$ is a *measure-preserving transformation* if, for all $A \in \mathscr{B}$, $T^{-1}(A) \in \mathscr{B}$ and $\mu(T^{-1}(A)) = \mu(A)$. The tuple $(X, \mathscr{B}, \mu, T)$ is called a *measure-preserving dynamical system*. A measure-preserving dynamical system $(Y, \mathscr{C}, \nu, S)$ is a *factor* of $(X, \mathscr{B}, \mu, T)$ if there is a function $\pi : X \to Y$ such that, for $A \subseteq Y$, $A \in \mathscr{C}$ if and only if $\pi^{-1}(A) \in \mathscr{B}$, $\nu = \pi_* \mu$—which means $\nu(A) = \mu\left(\pi^{-1}(A)\right)$ for every $A \in \mathscr{C}$—and $(S \circ \pi)(x) = (\pi \circ T)(x)$ for $\mu$-almost every $x \in X$.

*Example 5.23* Suppose that $X = [0, N-1]$ is an infinite hyperfinite interval. Define $S : X \to X$ by $S(x) = x + 1$ if $x < N$ and $S(N-1) = 0$. Then $S$ is a measure-preserving transformation and the dynamical system $(X, \mathscr{L}_X, \mu_X, S)$ will be referred to as a *hypercycle system*.

The hypercycle system will play an important role later in the book. In particular, we will need to use the *pointwise ergodic theorem* for the hypercycle system. While the proof of the general ergodic theorem is fairly nontrivial, the proof for the hypercycle system, due to Kamae [81], is much simpler. In the rest of this section, we fix a hypercycle system $(X, \Omega_X, \mu_X, S)$.

**Theorem 5.24 (The Ergodic Theorem for the Hypercycle System)** *Suppose that* $f \in L^1(X, \Omega, \mu)$. *Define*

$$\hat{f}(x) := \lim_{n \to \infty} \frac{1}{n} \sum_{i=0}^{n-1} f(S^i x)$$

*whenever this limit exists. Then:*

1. $\hat{f}(x)$ *exists for almost all* $x \in X$;
2. $\hat{f} \in L^1(X, \Omega, \mu)$;
3. $\int_X f d\mu = \int_X \hat{f} d\mu$.

*Proof* Without loss of generality, we may assume that $X = [0, N - 1]$ for some $N > \mathbb{N}$ and $f(x) \geq 0$ for $\mu_X$-almost every $x \in X$. We set

$$\overline{f}(x) := \limsup_{n \to \infty} \frac{1}{n} \sum_{i=0}^{n-1} f(S^i x)$$

and

$$\underline{f}(x) := \liminf_{n \to \infty} \frac{1}{n} \sum_{i=0}^{n-1} f(S^i x).$$

Note that $\overline{f}, \underline{f}$ are $\mu_X$-measurable and $S$-invariant. It suffices to show that $\overline{f}, \underline{f} \in L^1(X, \Omega, \mu)$ and that

$$\int_X \overline{f} d\mu \leq \int_X f d\mu \leq \int_X \underline{f} d\mu.$$

Towards this end, fix $\epsilon > 0$ and $m \in \mathbb{N}$. By Lemma 5.19, we may find internal functions $F, G : [0, N - 1] \to {}^*\mathbb{R}$ such that:

- for all $x \in X$, we have $f(x) \leq F(x)$ and $G(x) \leq \min\{\overline{f}(x), m\}$;
- for every internal subset $B$ of $X$

$$\max\left\{\left|\int_B f d\mu - \frac{1}{N} \sum_{x \in B} F(x)\right|, \left|\int_B \min\{\overline{f}, m\} d\mu - \frac{1}{N} \sum_{x \in B} G(x)\right|\right\} < \epsilon.$$

By definition of $\overline{f}$, for each $x \in X$, there is $n \in \mathbb{N}$ such that $\min\{\overline{f}(x), m\} \leq \frac{1}{n} \sum_{i=0}^{n-1} f(S^i x) + \epsilon$. For such an $n$ and $k = 0, 1, \ldots, n - 1$, we then have that

$$G(S^k x) \leq \min\{\overline{f}(S^k x), m\} = \min\{\overline{f}(x), m\} \leq \frac{1}{n} \sum_{i=0}^{n-1} f(S^i x) + \epsilon$$

$$\leq \frac{1}{n} \sum_{i=0}^{n-1} F(S^i x) + \epsilon,$$

whence it follows that

$$\sum_{i=0}^{n-1} G(S^i x) \leq \sum_{i=0}^{n-1} F(S^i x) + n\epsilon. \tag{5.1}$$

Since the condition in (5.1) is internal, the function $\rho : X \to {}^*\mathbb{N}$ that sends $x$ to the least $n$ making (5.1) hold for $x$ is internal. Note that $\rho(x) \in \mathbb{N}$ for all $x \in K$, whence $\sigma := \max_{x \in X} \rho(x) \in \mathbb{N}$.

Now one can start computing the sum $\sum_{x=0}^{N} G(x)$ by first computing

$$\sum_{x=0}^{\rho(0)-1} G(x) = \sum_{x=0}^{\rho(0)-1} G(S^x 0),$$

which is the kind of sum appearing in (5.1). Now in order to continue the computation using sums in which (5.1) applies, we next note that

$$\sum_{x=\rho(0)}^{\rho(0)+\rho(\rho(0))-1} G(x) = \sum_{x=0}^{\rho(\rho(0))-1} G(S^x \rho(0)).$$

This leads us to define, by internal recursion (which is the statement obtained from the usual principle of induction by applying transfer), the internal sequence $(\ell_j)$ by declaring $\ell_0 := 0$ and $\ell_{j+1} := \ell_j + \rho(\ell_j)$. It follows that we have

$$\sum_{x=0}^{\ell_J-1} G(x) = \sum_{j=0}^{J-1} \sum_{i=0}^{\rho(\ell_j)-1} G(S^i \rho(\ell_j)) \le \sum_{j=0}^{J-1} \sum_{j=0}^{\rho(\ell_j)-1} F(S^i x) + \rho(\ell_j)\epsilon$$

$$= \sum_{x=0}^{\ell_J-1} F(x) + \ell_J \epsilon.$$

As a result, we have that, whenever $\ell_J < N$,

$$\frac{1}{N} \sum_{x=0}^{\ell_J-1} G(x) \le \frac{1}{N} \sum_{x=0}^{\ell_J-1} F(x) + \epsilon.$$

Now take $J$ such that $N - \sigma \le \ell_J < N$. Since $\sigma \in \mathbb{N}$ and $G(x) \le m$ for every $x \in X$, we have that

$$\int_X \min\{\bar{f}, m\} d\mu \le \frac{1}{N} \sum_{x=0}^{N-1} G(x) + \epsilon \approx \frac{1}{N} \sum_{x=0}^{\ell_J-1} G(x) + \epsilon$$

$$\le \frac{1}{N} \sum_{x=0}^{\ell_J-1} F(x) + 2\epsilon \approx \frac{1}{N} \sum_{x=0}^{N-1} F(x) + 2\epsilon \le \int_X f d\mu + 3\epsilon.$$

Letting $m \to \infty$ and then $\epsilon \to 0$, we get that $\bar{f} \in L^1(X, \Omega, \mu)$ and $\int_X \bar{f} d\mu \le \int_X f d\mu$. The inequality $\int_X f d\mu \le \int_X \underline{f} d\mu$ is proven similarly.

In [81], Kamae uses the previous theorem to prove the ergodic theorem for an arbitrary measure-preserving dynamical system. In order to accomplish this, he proves the following result, which is interesting in its own right.

**Theorem 5.25 (Universality of the Hypercycle System)** *Suppose that $(Y, \mathscr{B}, \nu)$ is a standard probability space[1] and $T : Y \to Y$ is an measure-preserving transformation. Then $(Y, \mathscr{B}, \nu, T)$ is a factor of the hypercycle system $(X, \Omega_X, \mu_X, S)$.*

*Proof* As before, we may assume that $X = [0, N - 1]$ for some $N > \mathbb{N}$. Without loss of generality, we can assume that $(Y, \mathscr{B}, \nu)$ is atomless, and hence isomorphic to $[0, 1]$ endowed with the Borel $\sigma$-algebra and the Lebesgue measure. Consider the Borel map $r : [0, 1] \to [0, 1]^{\mathbb{N}}$ given by $r(y)(n) = h(T^n y)$ and the measure $r_* \nu$ on the Borel $\sigma$-algebra of $[0, 1]^{\mathbb{N}}$. Then $r$ defines an isomorphism between $(Y, \mathscr{B}, \nu, T)$ and a factor of the unilateral Bernoulli shift on $[0, 1]^{\mathbb{N}}$. Therefore, it is enough to consider the case when $(Y, \mathscr{B}, \nu, T)$ is the unilateral Bernoulli shift on $[0, 1]^{\mathbb{N}}$ endowed with the Borel $\sigma$-algebra $\mathscr{B}$ and some shift-invariant Borel probability measure $\nu$.

We now define the factor map $\pi : X \to [0, 1]^{\mathbb{N}}$. In order to do this, we fix $\alpha \in [0, 1]^{\mathbb{N}}$ such that $\lim_{n \to \infty} \frac{1}{n} \sum_{i=0}^{n-1} f(T^i \alpha) = \int_{[0,1]^{\mathbb{N}}} f(y) d\nu$ for all $f \in C([0, 1]^{\mathbb{N}})$. Such an $\alpha$ is called *typical* in [81] and is well-known to exist.[2]

By transfer, one can identify $^*([0, 1]^{\mathbb{N}})$ with the set of internal functions from $^*\mathbb{N}$ to $^*[0, 1]$. By compactness of $[0, 1]^{\mathbb{N}}$, one can deduce that, given $\xi \in {}^*([0, 1]^{\mathbb{N}})$, there exists a unique element $\mathrm{st}(\xi) \in [0, 1]^{\mathbb{N}}$ such that $\xi \approx \mathrm{st}(\xi)$, in the sense that, for every open subset $U$ of $[0, 1]^{\mathbb{N}}$, one has that $\xi \in {}^*U$ if and only if $\mathrm{st}(\xi) \in {}^*U$. (This also follows from Example 2.76 and compactness of $[0, 1]^{\mathbb{N}}$.) Concretely, one can identify $\mathrm{st}(\xi)$ with the element of $[0, 1]^{\mathbb{N}}$ such that $\mathrm{st}(\xi)(n) = \mathrm{st}(\xi(n))$ for $n \in \mathbb{N}$.

The function $\mathbb{N} \to [0, 1]^{\mathbb{N}}$, $n \mapsto T^n \alpha$ has a nonstandard extension $^*\mathbb{N} \to {}^*([0, 1]^{\mathbb{N}})$. Given $i \in [0, N - 1]$, define $\pi(i) := \mathrm{st}(T^i \alpha)$. We must show that $\pi_* \mu_X = \nu$ and that $(T \circ \pi)(i) = (\pi \circ S)(i)$ for $\mu_X$-almost every $i \in [0, N - 1]$. For $f \in C([0, 1]^{\mathbb{N}})$, we have that

$$\int_{[0,1]^{\mathbb{N}}} f(y) d\nu = \lim_{n \to \infty} \frac{1}{n} \sum_{i=0}^{n-1} f(T^i \alpha) \approx \frac{1}{N} \sum_{i=0}^{N-1} f\left(T^i \alpha\right) \approx \int_X (f \circ \pi) d\mu_X.$$

---

[1]Unfortunately, *standard* is used in a different sense than in the rest of this book. Indeed, here, a standard probability space is simply a probability space which is isomorphic to a quotient of $[0, 1]$ endowed with the Borel $\sigma$-algebra and Lebesgue measure.

[2]Of course, one can use the ergodic theorem to prove the existence of typical elements. However, we need a proof that typical elements exist that does not use the ergodic theorem. One can see, for example, [81, Lemma 2] for such a proof.

Note that the first step uses the fact that $\alpha$ is typical and the last step uses the fact that $f$ is continuous and Theorem 5.18. This shows that

$$\int_{[0,1]^{\mathbb{N}}} f \, dv = \int_X (f \circ \pi) \, d\mu_X = \int_{[0,1]^{\mathbb{N}}} f \, d\pi_* \mu_X$$

and hence $v = \pi_* \mu_X$.

To finish, we show that $(T \circ \pi)(i) = (\pi \circ S)(i)$ for $\mu_X$-almost every $i \in X$. Fix $i \in [0, N-2]$. Then we have

$$T(\pi(i)) = T(\mathrm{st}(T^i \alpha)) = \mathrm{st}(T^{i+1} \alpha)) = \pi(S(i)),$$

where the second equality uses the fact that $T$ is continuous.

From Theorems 5.24 and 5.25, we now have a proof of the ergodic theorem for measure-preserving systems based on standard probability spaces. It only requires one more step to obtain the ergodic theorem in general.

**Corollary 5.26 (The Ergodic Theorem)** *Suppose that* $(Y, \mathscr{B}, v, T)$ *is a measure-preserving dynamical system and* $f \in L^1(X, \Omega, \mu)$. *Define* $\hat{f}(x) := \lim_{n \to \infty} \frac{1}{n} \sum_{i=0}^{n-1} f(T^i x)$ *whenever this limit exists. Then:*

1. $\hat{f}(x)$ *exists for almost all* $x \in Y$;
2. $\hat{f} \in L^1(Y, \mathscr{B}, v)$;
3. $\int_Y f \, dv = \int_Y \hat{f} \, dv$.

*Proof* Let $\tau : Y \to \mathbb{R}^{\mathbb{N}}$ be given by $\tau(y)(n) := f(T^n y)$. Let $\mathscr{C}$ denote the Borel $\sigma$-algebra of $\mathbb{R}^{\mathbb{N}}$. Let $\sigma$ be the shift operator on $\mathbb{R}^{\mathbb{N}}$. Let $g : \mathbb{R}^{\mathbb{N}} \to \mathbb{R}$ be given by $g(\alpha) = \alpha(0)$. It is then readily verified that the ergodic theorem for $(Y, \mathscr{B}, v, T, f)$ is equivalent to the ergodic theorem for $(\mathbb{R}^{\mathbb{N}}, \mathscr{C}, \tau_* v, \sigma, g)$, which, as we mentioned above, follows from Theorems 5.24 and 5.25.

## Notes and References

The Loeb measure construction was introduced by Loeb in 1973 [92]. The Loeb measure plays a crucial role in several applications of nonstandard methods to a wide variety of areas of mathematics, including measure theory, probability theory, and analysis. A survey of such applications can be found in [32]. The nonstandard proof of the ergodic theorem due to Kamae [81, 82] is just a single but insightful example of the usefulness of Loeb measure. The Loeb measure also underpins the nonstandard perspective on the Furstenberg correspondence theorem, which in turns opens the gates to application of nonstandard methods to additive number theory (see Chap. 10).

# Part II
# Ramsey Theory

# Chapter 6
# Ramsey's Theorem

Ramsey theory studies, generally speaking, the following problem: Suppose that a given structure is colored using finitely many colors (equivalently, partition into finitely many pieces). Which combinatorial configurations can be found that are monochromatic, i.e. consisting of elements of the same color (equivalently, entirely contained in one of the pieces)? Ramsey's theorem from 1930, which we will present in this chapter, can be seen as the foundational result in this area. While remarkably simple to state, it has a large number of important consequences and applications. Many of these applications were studied by Erdős and Rado in the 1950s, who "rediscovered" Ramsey's theorem and recognized it importance. Attempts to generalize Ramsey's theorem in different contexts and directions have been one of the main driving forces in Ramsey theory.

## 6.1   Infinite Ramsey's Theorem

Recall that a *graph* is a pair $(V, E)$ where $V$ is the set of *vertices*, and the set of *edges* $E \subseteq V \times V$ is an anti-reflexive and symmetric binary relation on $V$. If $X \subseteq V$ is such that $(x, x') \in E$ (resp. $(x, x') \notin E$) for all distinct $x, x' \in X$, we say that $X$ is a *clique* (resp. *anticlique*) in $(V, E)$.

**Theorem 6.1 (Ramsey's Theorem for Pairs)** *If $(V, E)$ is an infinite graph, then $(V, E)$ either contains an infinite clique or an infinite anticlique.*

*Proof* Let $\xi$ be an element of $^*V$ that does not belong to $V$. Consider the element $(\xi, {}^*\xi) \in {}^{**}V$. There are now two possibilities: either $(\xi, {}^*\xi) \in {}^{**}E$ or $(\xi, {}^*\xi) \notin {}^{**}E$. We only treat the first case, the second case being entirely similar. We recursively define a one-to-one sequence $(x_n)$ in $V$ such that the set $\{x_n : n \in \mathbb{N}\}$

© Springer Nature Switzerland AG 2019

M. Di Nasso et al., *Nonstandard Methods in Ramsey Theory and Combinatorial Number Theory*, Lecture Notes in Mathematics 2239, https://doi.org/10.1007/978-3-030-17956-4_6

forms a clique in $(V, E)$. Towards this end, suppose that $d \in \mathbb{N}$ and $x_0, \ldots, x_{d-1}$ are distinct elements of $V$ such that, for all $1 \le i < j < d$, we have

- $(x_i, x_j) \in E$, and
- $(x_i, \xi) \in {}^*E$.

Consider now the statement "there exists $y \in {}^*V$ such that, for $i < d$, $y$ is different from $x_i$, and $(x_i, y) \in {}^*E$, and $(y, {}^*\xi) \in {}^{**}E$", whose truth is witnessed by $\xi$. It follows by transfer that there exists $x_d \in V$ different from $x_i$ for $i < d$, such that $(x_i, x_d) \in E$ for $i < d$, and $(x_d, \xi) \in {}^*E$. This concludes the recursive construction.

In order to state the full Ramsey theorem, we need the notion of a hypergraph. Given $m \in \mathbb{N}$, an *m-regular hypergraph* is a set $V$ of vertices together with a subset $E$ of $V^m$ that is permutation-invariant and has the property that $(x_1, \ldots, x_m) \in E$ implies that $x_1, \ldots, x_m$ are pairwise distinct. A *clique* (resp. *anticlique*) for $(V, E)$ is a subset $Y$ of $V$ with the property that $(y_1, \ldots, y_m) \in E$ (resp. $(y_1, \ldots, y_m) \notin E$) for any choice of pairwise distinct elements $y_1, \ldots, y_m$ of $Y$.

**Theorem 6.2 (Ramsey's Theorem)** *If $(V, E)$ is an infinite m-regular hypergraph, then $(V, E)$ contains an infinite clique or an infinite anticlique.*

**Exercise 6.3** Prove Theorem 6.2.

Ramsey's theorem is often stated in the language of colorings. Given a set $X$ and $m \in \mathbb{N}$, we let $X^{[m]}$ denote the set of $m$-element subsets of $X$. If $X \subseteq \mathbb{N}$, we often identify $X^{[m]}$ with the set of pairs $\{(x_1, \ldots, x_m) \in X^m : x_1 < \cdots < x_m\}$. Given $k \in \mathbb{N}$, a *k-coloring* of $X^{[m]}$ is a function $c : X^{[m]} \to \{1, \ldots, k\}$. In this vein, we often refer to the elements of $\{1, \ldots, k\}$ as colors. Finally, a subset $Y \subseteq X$ is *monochromatic for the coloring c* if the restriction of $c$ to $Y^{[m]}$ is constant. Here is the statement of Ramsey's theorem for colorings.

**Corollary 6.4** *For any $k, m \in \mathbb{N}$, any infinite set $V$, and any k-coloring $c$ of $V^{[m]}$, there is an infinite subset of $V$ that is monochromatic for the coloring c.*

*Proof* By induction, it suffices to consider the case $k = 2$. We identify a coloring $c : V^{[m]} \to \{1, 2\}$ with the $m$-regular hypergraph $(V, E)$ satisfying $(x_1, \ldots, x_m) \in E$ if and only if $c(\{x_1, \ldots, x_m\}) = 1$ for distinct $x_1, \ldots, x_m \in V$. An infinite clique (resp. anticlique) in $(V, E)$ corresponds to an infinite set with color 1 (resp. 2), whence the corollary is merely a restatement of our earlier version of Ramsey's theorem.

*Remark 6.5* Ramsey's Theorem cannot be extended to finite colorings of the infinite parts $V^{[\infty]} = \{A \subseteq V \mid A \text{ is infinite}\}$. Indeed, pick a copy of the natural numbers $\mathbb{N} \subseteq V$, pick an infinite $\alpha \in {}^*\mathbb{N} \setminus \mathbb{N}$, and for $A \in V^{[\infty]}$ set $c(A) = 1$ if the internal cardinality $|{}^*A \cap [1, \alpha]|$ is odd, and $c(A) = 2$ otherwise. Then $c : V^{[\infty]} \to \{1, 2\}$ is a 2-coloring with the property that $X^{[\infty]}$ is *not* monochromatic for any infinite $X \subseteq V$ since, e.g., $c(X) \ne c(X \setminus \{x\})$ for every $x \in X$.

## 6.2 Finite Ramsey Theorem

Corollary 6.4 is often referred to as the infinite Ramsey theorem. We now deduce from it the finite Ramsey theorem. We first need a bit of notation.

**Definition 6.6** Given $k, l, m, n \in \mathbb{N}$, we write $l \to (n)_k^m$ if every coloring of $[l]^{[m]}$ with $k$ colors has a homogeneous set of size $n$.

**Corollary 6.7 (Finite Ramsey Theorem)** *For every* $k, m, n \in \mathbb{N}$, *there is* $l \in \mathbb{N}$ *such that* $l \to (k)_m^n$.

*Proof* Suppose the theorem is false for a particular choice of $k, m, n$. Then for every $l \in \mathbb{N}$, there is a "bad" coloring $c_l : [l]^{[m]} \to \{1, \ldots, k\}$ with no monochromatic subset of size $n$. We can form a finitely branching tree of bad colorings with the partial order being inclusion. Since there is a bad coloring for every such $l$, we have that the tree is infinite. By König's Lemma, there is an infinite branch. This branch corresponds to a coloring of $\mathbb{N}^{[m]} \to \{1, \ldots, k\}$ with no monochromatic subset of size $n$, contradicting the Infinite Ramsey Theorem.

The proof of Theorem 6.7 is a typical example of a *compactness argument*. Compactness arguments are often used in combinatorics to deduce from an infinitary combinatorial statement a corresponding finitary analogue. Implicitly, a compactness argument hinges on compactness of a suitable topological space, which in the case of Theorem 6.7 is $\prod_k n_k$ for a sequence $(n_k)$ in $\mathbb{N}$.

Topological compactness is intimately connected with ultrafilters, nonstandard methods, and compactness in first order logic. Indeed, we used topological compactness to prove the existence of nonprincipal ultrafilters. In turn, nonprincipal ultrafilters can be used, via the ultraproduct construction, to prove the compactness principle in first order logic (and the existence of nonstandard maps, as we have seen). These ties between different instances of compactness are reflected in the various ways one can formulate compactness arguments in combinatorics. For instance, in the case of Theorem 6.7, one could also conclude the argument by fixing a nonprincipal ultrafilter $\mathscr{U}$ on $\mathbb{N}$ and then setting $c(a) = \lim_{l, \mathscr{U}} c_l(a)$ for $a \in \mathbb{N}^{[m]}$. Equivalently, one can fix an infinite $L \in {}^*\mathbb{N}$ and then let $c$ be the restriction of $c_L$ to $\mathbb{N}^{[m]} \subset {}^*\mathbb{N}^{[m]}$.

## 6.3 Rado's Path Decomposition Theorem

In this section, by a *path in* $\mathbb{N}$ we mean a (finite or infinite) injective sequence of natural numbers. For a finite path $(a_0, \ldots, a_n)$ from $\mathbb{N}$, we refer to $a_n$ as the *end of the path*.

Suppose that $c : \mathbb{N}^{[2]} \to \{1, \ldots, r\}$ is an $r$-coloring of $\mathbb{N}^{[2]}$. For $i \in \{1, \ldots, r\}$, we say that a path $P = (a_n)$ has color $i$ if $c(\{a_n, a_{n+1}\}) = i$ for all $n$.

**Theorem 6.8 (Rado's Path Decomposition Theorem)** *Suppose that* $c : \mathbb{N}^{[2]} \to \{1, \ldots, r\}$ *is an $r$-coloring of* $\mathbb{N}^{[2]}$. *Then there is a partition of* $\mathbb{N}$ *into paths* $P_1, \ldots, P_r$ *such that each $P_i$ has color $i$.*

*Proof* First, fix $\alpha \in {}^*\mathbb{N}$. For $m \in \mathbb{N}$ and $i \in \{1, \ldots, r\}$, we say that $m$ has color $i$ if $c(\{m, \alpha\}) = i$. We now recursively define disjoint finite paths $P_{1,k}, \ldots, P_{r,k}$ such that, whenever $P_{i,k} \neq \emptyset$, then the end of $P_{i,k}$ has color $i$ (in the sense of the previous sentence).

To start, we define $P_{i,0} = \emptyset$ for each $i = 1, \ldots, r$. Now assume that $P_{i,k-1}$ has been constructed for $i = 1, \ldots, r$. If $k$ belongs to some $P_{i,k-1}$, then set $P_{i,k} := P_{i,k-1}$ for all $i = 1, \ldots, r$. Otherwise, let $i$ be the color of $k$ and let $e$ be the end of $P_{i,k-1}$. Since $c(\{k, \alpha\}) = c(\{e, \alpha\}) = i$, by transfer, we can find $f \in \mathbb{N}$ larger than all numbers appearing in $\bigcup_{i=1}^{r} P_{i,k-1}$ such that $c(\{k, f\}) = c(\{e, f\}) = i$. We then set $P_{j,k} := P_{j,k-1}$ for $j \neq i$ and $P_{i,k} := P_{i,k-1}{}^\frown(f, k)$. Note that the recursive assumptions remain true.

For $i = 1, \ldots, r$, we now set $P_i$ to be the union of $P_{i,k}$ for $k \in \mathbb{N}$. It is clear that $P_1, \ldots, P_r$ are as desired.

## 6.4  Ultrafilter Trees

Given a set $X$, we let $X^{[<\infty]}$ (resp. $X^{[\infty]}$) denote the set of finite (resp. infinite) subsets of $X$. Given $s \in \mathbb{N}^{[<\infty]}$ and $X \subseteq \mathbb{N}$, we say that $s$ is an *initial segment* of $X$, denoted $s \sqsubseteq X$, if there is $i \in \mathbb{N}$ such that $s = \{j \in X : j \leq i\}$.

**Definition 6.9** A subset $T$ of $\mathbb{N}^{[<\infty]}$ is called a *tree on* $\mathbb{N}$ if $T \neq \emptyset$ and for all $s, t \in \mathbb{N}^{[<\infty]}$, if $s \sqsubseteq t$ and $t \in T$, then $s \in T$.

For a tree $T$ on $\mathbb{N}$, we set

$$[T] := \{X \in \mathbb{N}^{[\infty]} : \forall s \in \mathbb{N}^{[<\infty]}(s \sqsubseteq X \Rightarrow s \in T)\}.$$

If there is an element of $T$ that is $\sqsubseteq$-maximal with respect to the property that it is $\sqsubseteq$-comparable to every element of $T$, we call this (necessarily unique) element of $T$ the *stem* of $T$, denoted stem($T$). Finally, given $s \in T$, we set $T/s := \{t \in T : s \sqsubseteq t\}$.

**Definition 6.10** Let $\mathcal{U} = \langle \mathcal{U}_s : s \in \mathbb{N}^{[<\infty]} \rangle$ be a family of nonprincipal ultrafilters on $\mathbb{N}$ and let $T$ be a tree on $\mathbb{N}$. We say that $T$ is a $\mathcal{U}$-*tree* if it has a stem stem($T$), $T/\text{stem}(T)$ is nonempty, and for all $s \in T/\text{stem}(T)$, we have that $\{n \in \mathbb{N} : s \cup \{n\} \in T\} \in \mathcal{U}_s$.

Note that a $\mathcal{U}$-tree $T$ contains no $\sqsubseteq$-maximal elements and that, for every $s \in T$, there is $X \in [T]$ such that $s \sqsubseteq X$.

The goal of this section is to prove the following Ramsey-theoretic statement about ultrafilter trees, recently proven by Trujillo in [126]:

**Theorem 6.11** *Suppose that $\mathscr{U} = \langle \mathscr{U}_s : s \in \mathbb{N}^{[<\infty]} \rangle$ is a sequence of non-principal ultrafilters on $\mathbb{N}$, $T$ is a $\mathscr{U}$-tree on $\mathbb{N}$, and $\mathscr{X} \subseteq \mathbb{N}^{[\infty]}$. Then there is a $\mathscr{U}$-tree $S \subseteq T$ with $\mathrm{stem}(S) = \mathrm{stem}(T)$ such that one of the following holds:*

1. $[S] \subseteq \mathscr{X}$;
2. $[S] \cap \mathscr{X} = \emptyset$;
3. *for every $\mathscr{U}$-tree $S'$ with $S' \subseteq S$, we have $[S'] \not\subseteq \mathscr{X}$ and $[S'] \cap \mathscr{X} \neq \emptyset$.*

Using hyperfinite generators of ultrafilters, we obtain the following nonstandard analogue of $\mathscr{U}$-trees:

**Definition 6.12** Let $\alpha = \langle \alpha_s : s \in \mathbb{N}^{[<\infty]} \rangle$ be a family of infinite elements of $^*\mathbb{N}$ and let $T$ be a tree on $\mathbb{N}$. We say that $T$ is a $\alpha$-*tree* if, for all $s \in T/\mathrm{stem}(T)$, we have that $s \cup \{\alpha_s\} \in {}^*T$.

Before proving Theorem 6.11, we need one key lemma:

**Lemma 6.13** *Fix $\alpha = \langle \alpha_s : s \in \mathbb{N}^{[<\infty]} \rangle$ with each $\alpha_s$ infinite. Suppose that $C \subseteq \mathbb{N}^{[<\infty]}$ is such that, for all $s \in C$, we have that $s \cup \{\alpha_s\} \in {}^*C$. Then for all $\alpha$-trees $T$, if $\mathrm{stem}(T) \in C$, then there is a $\alpha$-tree $S \subseteq T$ with $\mathrm{stem}(S) = \mathrm{stem}(T)$ such that $S/\mathrm{stem}(S) \subseteq C$.*

*Proof* Suppose that $T$ is a $\alpha$-tree with $\mathrm{stem}(T) \in H$. We first recursively define sets $L_n \subseteq C \cap T$ as follows. Set $L_0 := \{\mathrm{stem}(T)\}$. Supposing that $L_n$ has been defined, we set

$$L_{n+1} := \{s \cup \{m\} : s \in L_n, \ m > \max(s), \text{ and } s \cup \{m\} \in C \cap T\}.$$

We now set

$$S := \{s \in \mathbb{N}^{[<\infty]} : s \sqsubseteq \mathrm{stem}(T)\} \cup \bigcup_{n=0}^{\infty} L_n.$$

We claim that this $S$ is as desired. It follows directly by induction that $S$ is a tree on $\mathbb{N}$ and that $S \subseteq T$. Moreover, by the hypothesis on $C$ and the fact that $T$ is an $\alpha$-tree, we have that $S$ is also an $\alpha$-tree. It is clear that $\mathrm{stem}(T) \sqsubseteq \mathrm{stem}(S)$. However, since $\mathrm{stem}(T) \cup \{\alpha_{\mathrm{stem}(T)}\} \in {}^*L_1$, we have that $\{n \in \mathbb{N} \ \mathrm{stem}(T) \cup \{n\} \in L_1\}$ is infinite, whence it follows that $\mathrm{stem}(S) = \mathrm{stem}(T)$. Finally, $S/\mathrm{stem}(S) = \bigcup_{n=0}^{\infty} L_n \subseteq C$.

We can now prove Theorem 6.11 in its equivalent nonstandard formulation:

**Theorem 6.14** *Suppose that $\alpha = \langle \alpha_s : s \in \mathbb{N}^{[<\infty]} \rangle$ is a sequence of infinite elements of $^*\mathbb{N}$, $T$ is a $\alpha$-tree on $\mathbb{N}$, and $\mathscr{X} \subseteq \mathbb{N}^{[\infty]}$. Then there is a $\alpha$-tree $S \subseteq T$ with $\mathrm{stem}(S) = \mathrm{stem}(T)$ such that one of the following holds:*

1. $[S] \subseteq \mathscr{X}$;
2. $[S] \cap \mathscr{X} = \emptyset$;
3. *for every $\alpha$-tree $S'$ with $S' \subseteq S$, we have $[S'] \not\subseteq \mathscr{X}$ and $[S'] \cap \mathscr{X} \neq \emptyset$.*

*Proof* We introduce the following three sets:

$$A := \{s \in \mathbb{N}^{[<\infty]} \; : \; \text{there is a } \alpha\text{-tree } S \subseteq T \text{ with stem}(S) = s \text{ and } [S] \subseteq \mathcal{X}\},$$

$$B := \{s \in \mathbb{N}^{[<\infty]} \; : \; \text{there is a } \alpha\text{-tree } S \subseteq T \text{ with stem}(S) = s \text{ and } [S] \subseteq \mathbb{N}^{\infty} \setminus \mathcal{X}\},$$

$$C := \mathbb{N}^{[<\infty]} \setminus (A \cup B).$$

*Claim* If $s \in C$, then $s \cup \{\alpha_s\} \in {}^{*}C$.

*Proof of Claim* We argue by contrapositive, whence we assume that $s \cup \{\alpha_s\} \in {}^{*}A \cup {}^{*}B$. We only treat the case that $s \cup \{\alpha_s\} \in {}^{*}A$, the other case being similar. Let $D := \{n \in \mathbb{N} \; : \; s \cup \{n\} \in A\}$. Note that $\alpha_s \in {}^{*}D$. For each $n \in D$, let $T_n$ be a $\alpha$-tree with stem$(T_n) = s \cup \{n\}$ and $[T_n] \subseteq \mathcal{X}$. Let $S := \bigcup_{n \in D} T_n$. Observe that:

(i) $S$ is a tree,
(ii) stem$(S) = s$,
(iii) $\{s \cup \{n\} \; : \; n \in D\} \subseteq S$, and
(iv) $[S] = \bigcup_{n \in D} [T_n] \subseteq \mathcal{X}$.

It remains to show that $S$ is a $\alpha$-tree, for then $s \in A$, as desired. Thus, given $t \in S$, we need $t \cup \{\alpha_t\} \in {}^{*}S$. If $t = s$, then $s \cup \{\alpha_s\} \in {}^{*}S$ by item (iii) and the above observation that $\alpha_s \in {}^{*}D$. Otherwise, there is $n \in D$ such that $t \in T_n/(s \cup \{n\})$. Since $T_n$ is a $\alpha$-tree, we have that $t \cup \{\alpha_t\} \in {}^{*}T_n \subseteq {}^{*}S$. This finishes the proof of the claim.

It is clear that if stem$(T) \in A$ (resp. stem$(T) \in B$), then item (1) (resp. item (2)) of the conclusion of the theorem holds. We may thus suppose that stem$(T) \in C$. By Lemma 6.13, there is a $\alpha$-tree $S \subseteq T$ with stem$(S) = $ stem$(T)$ such that $S/\text{stem}(S) \subseteq C$. We claim that this $S$ is as desired. Indeed, suppose that $S'$ is a $\alpha$-tree with $S' \subseteq S$. Then stem$(S') \in S/\text{stem}(S) \subseteq C$. It follows from the definition of $C$ that $[S'] \not\subseteq \mathcal{X}$ and $[S'] \cap \mathcal{X} \neq \emptyset$, as desired.

We offer one application of Theorem 6.11. Given a tree $T$ on $\mathbb{N}$ and $n \in \mathbb{N}$, we set $T(n) := T \cap \mathbb{N}^{[n]}$.

**Corollary 6.15 (Ramsey's Theorem for $\mathcal{U}$-Trees)** *Fix $n \in \mathbb{N}$ and $A \subseteq \mathbb{N}^{[n]}$. Further fix a sequence $\mathcal{U} = \langle \mathcal{U}_s \; : \; s \in \mathbb{N}^{[<\infty]} \rangle$ of nonprincipal ultrafilters on $\mathbb{N}$ and a $\mathcal{U}$-tree $T$. Then there is a $\mathcal{U}$-tree $S \subseteq T$ with stem$(S) = $ stem$(T)$ such that either $S(n) \subseteq A$ or $S(n) \cap A = \emptyset$.*

*Proof* For each $Y \subseteq \mathbb{N}$ with $|Y| \geq n$, set $r_n(Y) \in \mathbb{N}^{[n]}$ to be the unique $s \in \mathbb{N}^{[n]}$ with $s \sqsubseteq Y$. Set $\mathcal{X} := \{Y \in \mathbb{N}^{[\infty]} \; : \; r_n(Y) \in A\}$. We apply Theorem 6.11 to $\mathcal{U}$, $T$, and $\mathcal{X}$, obtaining a $\mathcal{U}$-tree $S$ with $S \subseteq T$ and stem$(S) = $ stem$(T)$. Note that $S$ cannot satisfy item (3) in the conclusion of Theorem 6.11: if $S'$ is a $\mathcal{U}$-tree with $S' \subseteq S$ and $|$ stem$(S')| \geq n$, then either $r_n(\text{stem}(S')) \in A$ (whence $[S'] \subseteq \mathcal{X}$) or $r_n(\text{stem}(S')) \notin A$ (whence $[S'] \cap \mathcal{X} = \emptyset$). Consequently, either $[S] \subseteq \mathcal{X}$ (whence $S(n) \subseteq A$) or $[S] \cap \mathcal{X} = \emptyset$ (whence $S(n) \cap A = \emptyset$).

# Notes and References

The Ramsey theorem was proved in the foundational paper of Ramsey [111]. In fact, in this paper the theorem is obtained as an intermediate step towards establishing a result in propositional logic, hence the title "On a problem of formal logic". While Ramsey's theorem did not initially receive too much attention, it was later "rediscovered" in the 1950s by Erdős and Rado who recognized its fundamental importance and provided several variations and applications, such as Rado's decomposition theorem [110]. For more on the metamathematics of Rado's Decomposition Theorem, see [27], whose ultrafilter proof of the theorem is essentially the proof given here. Ultrafilter trees were first introduced by Blass in [17] and are part of the much larger *local Ramsey theory*, extensively developed in the book [124].

# Chapter 7
# The Theorems of van der Waerden and Hales-Jewett

As we have seen in the previous chapter, Ramsey's theorem proved to be crucial in the development of Ramsey theory, to which it gave its name. However, Ramsey theory has another equally important root in works of Hindman, Shur, and van der Waerden motivated by problems about rational functions and modular arithmetic: Hilbert's Cuble Lemma, Schur's Lemma, and van der Waerden's Theorem on arithmetic progressions. Particularly, the latter is of fundamental importance, as it paved the way to many of the later developments in Ramsey theory, including partition regularity of diophantine equations (see Chap. 9) and density results in additive combinatorics (see Chap. 10). The combinatorial essence of van der Waerden' theorem was later isolated by Hales and Jewett, who proved a powerful abstract pigeonhole principle, later generalized further by Graham and Rothschild. In this chapter, we will present nonstandard proofs of both van der Waerden's Theorem and the Hales–Jewett theorem.

## 7.1 The Theorem of van der Waerden

The van der Waerden theorem is one of the earliest achievements of what is now called Ramsey theory. Indeed, it was established by van der Waerden in 1928 [130], thus predating Ramsey's theorem itself. The theorem is concerned with the notion of *arithmetic progressions* in the set $\mathbb{N}$ of natural numbers. More precisely, for $k \in \mathbb{N}$, a $k$-term arithmetic progression in $\mathbb{N}$ is a set of the form $a + d[0, k) := \{a, a + d, a + 2d, \ldots, a + (k - 1)d\}$ for some $a, d \in \mathbb{N}$. A $k$-term arithmetic progression is also called an arithmetic progression of length $k$. An arithmetic progression in $^*\mathbb{N}$ is defined in a similar fashion, where one can actually consider $k$-term arithmetic progressions for $k \in {}^*\mathbb{N}$.

Recall that, for $k \in \mathbb{N}$, a $k$-*coloring* of a set $A$ is a function from $A$ to the set $[1, k] = \{1, \ldots, k\}$. A *finite coloring* of a $A$ is a $k$-coloring for some $k \in \mathbb{N}$.

© Springer Nature Switzerland AG 2019
M. Di Nasso et al., *Nonstandard Methods in Ramsey Theory and Combinatorial Number Theory*, Lecture Notes in Mathematics 2239, https://doi.org/10.1007/978-3-030-17956-4_7

A subset $B$ of $A$ is *monochromatic* with respect to a coloring $c$ if it is contained in the preimage of $i$ under $c$ for some $i \in [1, k]$. A collection $\mathscr{C}$ of subsets of $\mathbb{N}$ is *partition regular* if it is closed under supersets and, for any $A \in \mathscr{C}$ and finite coloring $c$ of $A$, there is a monochromatic $B \subseteq A$ such that $B \in \mathscr{C}$.

**Theorem 7.1** *The following are equivalent:*

1. *Every finite coloring of $\mathbb{N}$ admits arbitrarily long monochromatic arithmetic progressions.*
2. *For every $r, k \in \mathbb{N}$, there is $l \in \mathbb{N}$ such that every $r$-coloring of $[1, l]$ admits a monochromatic $k$-term arithmetic progression.*
3. *The property of containing arbitrarily long arithmetic progressions is partition regular.*

*Proof* (1)$\Rightarrow$(2) Suppose that (2) fails for some $k, r$. By overflow, there is $L > \mathbb{N}$ and an internal $r$-coloring of $[1, L]$ with no monochromatic $k$-term arithmetic progression. By considering the restriction of $c$ to $\mathbb{N}$, we get an $r$-coloring of $\mathbb{N}$ with no monochromatic $k$-term arithmetic progression, whence (1) fails.

(2)$\Rightarrow$(3) Suppose that (2) holds. Towards establishing (3), fix a set $A$ containing arbitrarily long arithmetic progressions and a partition of $A$ into two pieces $A = B_1 \sqcup B_2$. Fix $k \in \mathbb{N}$. Let $l$ witness the truth of (2) with two colors and $k$-term arithmetic progressions. Fix an arithmetic progression $x + [0, l)d \subseteq A$. For $i = 1, 2$, let $C_i := \{n \in [0, l) : x + nd \in B_i\}$. Then there is $i \in \{1, 2\}$ such that $C_i$ contains an arithmetic progression $y + [0, k)e$. It follows that $(x + yd) + [0, k)de$ is a $k$-term arithmetic progression contained in $B_i$. Since some $i$ must work for infinitely many $k$'s, we see that some $B_i$ contains arbitrarily long arithmetic progressions.

(3)$\Rightarrow$(1) This is obvious.

The following is a nonstandard presentation of the proof of van der Waerden's theorem from [60]; see also [106, Section 2.3]. First, some terminology. For $k, m \in \mathbb{N}$ and $g, h \in [0, k]^m$, we say that $g$ and $h$ are equivalent, written $g \equiv h$, if $g$ and $h$ agree up to the last occurrence of $k$.

**Definition 7.2** For $k, m \in \mathbb{N}$, let $S(k, m, r, n)$ be the statement: for any $r$-coloring of $[1, n]$, there exist $a, d_0, \ldots, d_{m-1} \in [1, n]$ such that $a + k \sum_{j < m} d_j \in [1, n]$ and, for any $g, h \in [0, k]^m$ such that $g \equiv h$, the elements $a + \sum_{j < m} g_j d_j$ and $a + \sum_{j < m} h_j d_j$ have the same color. We then let $S(k, m)$ be the statement: for all $r \in \mathbb{N}$, there is $n \in \mathbb{N}$ such that $S(k, m, r, n)$ holds.

We first observe that even though the statement $S(k, m, r, n)$ considers colorings of $[1, n]$, it is readily verified that its truth implies the corresponding statement for colorings of any interval of length $n$.

We next observe that the finitary van der Waerden theorem is the statement that $S(k, 1)$ holds for all $k \in \mathbb{N}$. Indeed, suppose that $S(k, 1)$ holds and fix $r \in \mathbb{N}$. Fix $n \in \mathbb{N}$ such that $S(k, 1, r, n)$ holds. Let $c : [1, n] \to [1, r]$ be an $r$-coloring of $[1, n]$. Then there is $a, d \in [1, n]$ such that $a + kd \in [1, n]$ and, since all elements of $[0, k]^1$ are equivalent, we get that $c(a + gd) = c(a + hd)$ for all $g, h \in [0, k]$, whence we get a monochromatic arithmetic progression of length $k + 1$.

If $v \in {}^*\mathbb{N}$, then we also consider the internal statement $S(k, m, r, v)$ which is defined exactly as its standard counterpart except that it only considers internal $r$-colorings of $[1, v]$.

**Lemma 7.3** $S(k, m)$ *is equivalent to the statement: for all* $r \in \mathbb{N}$ *and all* $v \in {}^*\mathbb{N} \backslash \mathbb{N}$, *we that have that* $S(k, m, r, v)$ *holds.*

*Proof* First suppose that $S(k, m)$ holds. Given $r \in \mathbb{N}$, take $n \in \mathbb{N}$ such that $S(k, m, r, n)$ holds. Fix $v \in {}^*\mathbb{N} \backslash \mathbb{N}$ and consider an internal $r$-coloring $c$ of $[1, v]$. Then $c|_{[1,n]}$ is an $r$-coloring of $[1, n]$, whence the validity of $S(k, m, r, n)$ yields the desired conclusion. Conversely, if $S(k, m, r, v)$ holds for all $v \in {}^*\mathbb{N} \backslash \mathbb{N}$, then by underflow there is $n \in \mathbb{N}$ such that $S(k, m, r, n)$ holds.

**Theorem 7.4** $S(k, m)$ *holds for all* $k, m \in \mathbb{N}$.

*Proof* Suppose, towards a contradiction, that $S(k, m)$ fails for the pair $(k, m)$ and that $(k, m)$ is lexicographically least with this property.

*Claim* $m = 1$.

*Proof of Claim* Suppose the claim is false. We obtain a contradiction by showing that $S(k, m, r, v)$ holds for all $r \in \mathbb{N}$ and all $v \in {}^*\mathbb{N} \backslash \mathbb{N}$. Towards this end, fix $r \in \mathbb{N}$, $v \in {}^*\mathbb{N} \backslash \mathbb{N}$, and an internal coloring $c : [1, v] \to [1, r]$. Since $S(k, m - 1)$ is true, there is $M \in \mathbb{N}$ such that $S(k, m - 1, r, M)$ is true. Write $v = NM + s$ with $0 \leq s < M$. Note that $N \in {}^*\mathbb{N} \backslash \mathbb{N}$. Consider the internal coloring $c_N : [1, N] \to [1, r^M]$ given by

$$c_N(i) := (c((i - 1)M + 1), \ldots, c((i - 1)M + M)).$$

Since $S(k, 1, r, N)$ holds, there is an arithmetic progression $b + d, b + 2d, \ldots, b + kd$ contained in $[1, N]$ that is monochromatic for the coloring $c_N$. Next, since $S(k, m - 1, r, M)$ holds, by considering $c|_{[(b-1)M, bM]}$, we see that there are $a, d_0, \ldots, d_{m-2} \in [(b - 1)M, bM]$ such that $a + k \sum_{j < m-1} d_j \in [(b - 1)M, bM]$ and, for any $g, h \in [0, k]^{m-1}$ such that $g \equiv h$, the elements $a + \sum_{j < m} g_j d_j$ and $a + \sum_{j < m} h_j d_j$ have the same color with respect to $c$.

Set $d_{m-1} := dM$. We claim that $a, d_0, \ldots, d_{m-1}$ are as desired. First note that $a + k \sum_{j < m} d_j \leq bM + kdM \leq NM \leq v$. Next suppose that $g, h \in [0, k]^m$ are such that $g \equiv h$. We wish to show that $a + \sum_{j < m} g_j d_j$ and $a + \sum_{j < m} h_j d_j$ have the same color. If the last occurrence of $k$ is $m - 1$, then this is obvious. Otherwise, we see that $g \mid m - 1 = h \mid m - 1$, whence by assumption $a + \sum_{j < m-1} g_j d_j$ and $a + \sum_{j < m-1} h_j d_j$ have the same color. Write $a + \sum_{j < m-1} g_j d_j = (b-1)M + p$ with $p \in [1, M]$. Then $a + \sum_{j < m} g_j d_j = (b - 1)M + p + g_{M-1} dM = (b + g_{m-1}d - 1)M + p$, which has the same color as $(b - 1)M + p$ by assumption. Likewise, $a + \sum_{j < m-1} h_j d_j = (b - 1)M + q$ with $q \in [1, M]$, whence $a + \sum_{j < m} h_j d_j = (b - 1)M + q + h_{M-1} dM = (b + h_{m-1}d - 1)M + q$, which has the same color as $(b - 1)M + q$ by assumption. Thus, $a + \sum_{j < m-1} g_j d_j$ and $a + \sum_{j < m-1} h_j d_j$ have the same color, proving the claim.

Since $S(k, 1)$ fails, necessarily we have $k > 1$. We will arrive at a contradiction by showing that $S(k, 1)$ in fact holds. Fix $r \in \mathbb{N}$, $\nu \in {}^*\mathbb{N}$ infinite, and an internal $r$-coloring $c$ of $[1, \nu]$. By minimality of $(k, 1)$, we have that there exist $a, d_0, \ldots d_{r-1} \in [1, \nu]$ such that $a + r \sum_{j<r} d_j \in [1, \nu]$ and, for any $g, h \in [1, k-1]^r$ with $g \equiv h$, we have $a + \sum_{j<r} g_j d_j$ and $a + \sum_{j<r} h_j d_j$ have the same color. Observe that there are $r + 1$ $r$-tuples that are obtained by concatenating a (possibly empty) $r$-tuple of $(k-1)$'s and a (possibly empty) $r$-tuple of $0$'s. Hence, by the pigeonhole principle, there exist $1 \leq s < t \leq r$ such that $a + (k-1) \sum_{i<s} d_i$ and $a + (k-1) \sum_{i<t} d_i$ have the same color. We also have that $a + (k-1) \sum_{i<s} d_i$ and $a + (k-1) \sum_{i<s} d_i + j \sum_{s \leq i < t} d_i$ have the same color for every $j < k - 1$. Therefore, setting $a' := a + (k-1) \sum_{i<s} d_i$ and $d' := \sum_{s \leq i < t} d_i$, we have that $a' + jd'$, for $j < k$, all have the same color. Since $\nu \in {}^*\mathbb{N}\backslash\mathbb{N}$ and $c$ were arbitrary, this witnesses that $S(k, 1)$ holds, yielding the desired contradiction.

We will see in the next section that the Hales-Jewett theorem allows us to immediately conclude a generalization of the van der Waerden theorem.

## 7.2 The Hales-Jewett Theorem

Let $L$ be a *finite* set (alphabet). We use the symbol $x$ to denote a *variable* not in $L$. We let $W_L$ denote the set of finite strings of elements of $L$ (called *words* in $L$), and $W_{Lx}$ denote the set of finite strings of elements of $L \cup \{x\}$ with the property that $x$ appears at least once (called *variable words*). We denote (variable) words by $v, w, z$ and letters by $a, b, c$. If $w$ is a variable word and $a$ is a letter, then we denote by $w[a]$ the word obtained from $w$ by replacing every occurrence of $x$ with $a$. For convenience, we also set $w[x] := w$. The concatenation of two (variable) words $v, w$ is denoted by $v^\frown w$.

**Definition 7.5** Fix a sequence $(w_n)$ of variable words

1. The *partial subsemigroup of* $W_L$ *generated by* $(w_n)$, denoted $[(w_n)]_{W_L}$, is the set of all words $w_{n_0}[a_0]^\frown \cdots ^\frown w_{n_{k-1}}[a_{k-1}]$, where $k \in \mathbb{N}$, $n_0 < \cdots < n_{k-1}$, and $a_0, \ldots, a_{k-1} \in L$.
2. The *partial subsemigroup of* $W_{Lx}$ *generated by* $(w_n)$, denoted $[(w_n)]_{W_{Lx}}$, is the set of all words $w_{n_0}[\lambda_0]^\frown \cdots ^\frown w_{n_{k-1}}[\lambda_{k-1}]$, where $k \in \mathbb{N}$, $n_0 < \cdots < n_{k-1}$, $\lambda_0, \ldots, \lambda_{k-1} \in L \cup \{x\}$, and some $\lambda_i = x$.

**Theorem 7.6 (Infinite Hales-Jewett)** *For every finite coloring of* $W_L \cup W_{Lx}$ *there exists an infinite sequence* $(w_n)$ *of variable words such that* $[(w_n)]_{W_L}$ *and* $[(w_n)]_{W_{Lx}}$ *are both monochromatic.*

There is also a finitary version of the Hales-Jewett theorem. Suppose that $x_1, \ldots, x_m$ are variables. A variable word $w$ in the variables $x_1, \ldots, x_m$ in the alphabet $L$ is a string of symbols in $L \cup \{x_1, \ldots, x_m\}$ such that, for every $1 \leq i \leq m$, $x_i$ occurs in $w$, and for every $1 \leq i < j \leq m$, the first occurrence of $x_i$ precedes the first occurrence of $x_j$. The word $w[a_1, \ldots, a_m]$ obtained from $w$ by substituting the variable $x_i$ with the letter $a_i$ for $i = 1, 2, \ldots, m$ is defined in the obvious way.

**Corollary 7.7 (Finite Hales-Jewett)** *For any finite alphabet $L$ and any $r, m \in \mathbb{N}$ there exists $n \in \mathbb{N}$ such that for any $r$-coloring of the set $W_L(n)$ of $L$-words of length $n$ there exist a variable word $w$ of length $n$ in the alphabet $L$ and variables $x_1, \ldots, x_m$ such that the "combinatorial $m$-subspace" $\{w[a_1, \ldots, a_m] : a_1, \ldots, a_n \in L\}$ is monochromatic.*

A combinatorial $m$-subspace for $m = 1$ is usually called a *combinatorial line*.

*Proof* We let $W_{Lx}(n)$ denote the elements of $W_{Lx}$ of length $n$ and $W_L(n)$ denote the elements of $W_L$ of length $n$. Suppose, towards a contradiction, that there is $r \in \mathbb{N}$ such that, for each $n$, there is a "bad" $r$-coloring of $W_L(n)$ that admits no monochromatic combinatorial line. By a compactness argument there is an $r$-coloring $c$ of $W_L$ such that the restriction of $c$ to $W_L(n)$ is a bad $r$-coloring for every $n \in \mathbb{N}$. By the Infinite Hales-Jewett Theorem, there is a sequence $(w_i)$ for which $[(w_i)]_{W_L}$ is monochromatic. For $i = 1, 2, \ldots, m$, rename the variable $x$ of $w_i$ by $x_i$, and consider the variable word $w := w_1 {}^\frown w_2 {}^\frown \cdots {}^\frown w_m$ in the variables $\{x_1, \ldots, x_m\}$. If $n$ is the length of $w$, then by the choice of $w_1, \ldots, w_m$ the combinatorial subspace $\{w[a_1, \ldots, a_m] : a_1, \ldots, a_n \in L\}$ is monochromatic. This contradicts the fact that the restriction of $c$ to $W_L(n)$ is a bad $r$-coloring. $\qquad\square$

From the Hales-Jewett theorem one can deduce a multidimensional generalization of van der Waerden's theorem, known as *Gallai's theorem*.

**Theorem 7.8 (Gallai)** *Fix $d \in \mathbb{N}$, a finite $F \subset \mathbb{N}^d$, and $r \in \mathbb{N}$. Then there exists $n \in \mathbb{N}$ such that, for any $r$-coloring of $[-n, n]^d$, there exist $\boldsymbol{a} \in \mathbb{N}^d$ and $c \in \mathbb{N}$ such that the affine image $\boldsymbol{a} + cF := \{\boldsymbol{a} + c\boldsymbol{x} : \boldsymbol{x} \in F\}$ of $F$ is monochromatic.*

*Proof* Consider the finite alphabet $L = F$. For $n \in \mathbb{N}$, consider the map $\Psi_n : W_L(n) \to \mathbb{N}^d$ defined by $\Psi_n((\boldsymbol{a}_1, \ldots, \boldsymbol{a}_n)) = \boldsymbol{a}_1 + \cdots + \boldsymbol{a}_n$. Observe that $\Psi_n$ maps a combinatorial line to an affine image of $F$. Thus the conclusion follows from the finitary Hales-Jewett theorem. $\qquad\square$

In the rest of the section we present the proof of Theorem 7.6. Consider $W_L$ and $W_L \cup W_{Lx}$ as semigroups with respect to concatenation. Thus their nonstandard extensions ${}^*W_L$ and ${}^*W_L \cup {}^*W_{Lx}$ have canonical semigroup operations with respect to the nonstandard extension of the concatenation operation, which we still denote by "$\frown$". The elements of ${}^*W_L$ can be regarded as hyperfinite strings of elements of ${}^*L$, and similarly for ${}^*W_{Lx}$. For every $a \in L \cup \{x\}$ we also denote by $\varpi \mapsto \varpi[a]$ the nonstandard extension of the substitution operation $W_{Lx} \to W_L$, $w \mapsto w[a]$.

**Lemma 7.9** *There exists a $u$-idempotent $\varpi$ in ${}^*W_{Lx}$ and a $u$-idempotent $\upsilon \in {}^*W_L$ such that $\varpi {}^\frown {}^*\upsilon \sim \upsilon {}^\frown {}^*\varpi \sim \varpi$ and $\varpi[a] \sim \upsilon$ for every $a \in L$.*

*Proof* Fix an enumeration $\{a_1, \ldots, a_m\}$ of $L$. We define, by recursion on $k = 1, \ldots, m$, $u$-idempotent elements $\varpi_1, \ldots, \varpi_m$ of $^*W_{Lx}$ and $\upsilon_1, \ldots, \upsilon_m$ of $^*W_L$ such that, for $1 \le i \le j \le m$,

1. $\varpi_j [a_i] \sim \upsilon_j$, and
2. $\varpi_j \sim \varpi_j {}^\frown{}^* \upsilon_i \sim \upsilon_i {}^\frown{}^* \varpi_j$.

Supposing this has been done, the conclusion of the lemma holds by taking $\varpi := \varpi_m$ and $\upsilon := \upsilon_m$.

To begin, we let $\varpi_0$ be any nontrivial $u$-idempotent element of $^*W_{Lx}$ and set $\upsilon_1 := \varpi_0 [a_1]$, which we note is an $u$-idempotent element of $^*W_L$. Let $\rho_1$ be an element of $^*W_{Lx}$ such that $\rho_1 \sim \varpi_0 {}^\frown{}^* \upsilon_1$. Observe that $\rho_1 [a_1] \sim \upsilon_1$ and $\rho_1 {}^\frown{}^* \upsilon_1 \sim \rho_1$. Thus, the compact $u$-semigroup

$$\left\{ z \in {}^*W_{Lx} : z [a_1] \sim \upsilon_1 \text{ and } z{}^\frown{}^* \upsilon_1 \sim z \right\}$$

is nonempty, whence it contains a $u$-idempotent $\beta_1$. We now fix $\varpi_1 \in {}^*W_{Lx}$ such that $\varpi_1 \sim \upsilon_1 {}^\frown{}^* \beta_1$. It follows now that $\varpi_1$ is $u$-idempotent and $\varpi_1$ and $\upsilon_1$ satisfy (1) and (2) above.

Suppose that $\varpi_i, \upsilon_i$ have been defined for $1 \le i \le k < m$ satisfying (1) and (2) above. Set $\upsilon_{k+1} := \varpi_k [a_{k+1}]$. Observe that $\upsilon_{k+1} \sim \upsilon_{k+1} {}^\frown{}^* \upsilon_i \sim \upsilon_i {}^\frown{}^* \upsilon_{k+1}$ for $1 \le i \le k+1$. Let $\rho_{k+1}$ be an element of $^*W_{Lx}$ such that $\rho_{k+1} \sim \varpi_k {}^\frown{}^* \upsilon_{k+1}$. Observe that $\upsilon_i {}^\frown{}^* \rho_{k+1} \sim \rho_{k+1} {}^\frown{}^* \upsilon_i \sim \rho_{k+1}$ and $\rho_{k+1} [a_i] \sim \upsilon_{k+1}$ for $1 \le i \le k+1$. Thus, the compact $u$-semigroup

$$\left\{ z \in {}^*W_{Lx} : z [a_i] \sim \upsilon_{k+1} \text{ and } z{}^\frown{}^* \upsilon_i \sim z \text{ for } 1 \le i \le k+1 \right\}$$

is nonempty, whence it contains a $u$-idempotent element $\beta_{k+1}$. Finally, fix $\varpi_{k+1}$ in $^*W_{Lx}$ such that $\varpi_{k+1} \sim \upsilon_{k+1} {}^\frown{}^* \beta_{k+1}$. It follows that $\varpi_{k+1}$ is $u$-idempotent and (1) and (2) continue to hold for $\varpi_i$ and $\upsilon_i$ for $1 \le i \le k+1$. This completes the recursive construction and the proof of the lemma.

In the statement of the following proposition, we assume that $\varpi$ and $\upsilon$ are as in the conclusion of Lemma 7.9.

**Proposition 7.10** *Suppose that $A \subset W_L$ and $B \subset W_{Lx}$ are such that $\upsilon \in {}^*A$ and $\varpi \in {}^*B$. Then there exists an infinite sequence $(w_n)$ in $W_{Lx}$ such that $[(w_n)]_{W_L}$ is contained in $A$ and $[(w_n)]_{W_{Lx}}$ is contained in $B$.*

*Proof* Set $C := A \cup B$. Observe that $\varpi$ satisfies, for every $a, b \in L \cup \{x\}$,

$$\varpi [a] \in {}^*C$$

$$\varpi [a] {}^\frown{}^* \varpi [b] \in {}^{**}C.$$

Therefore, by transfer, there exists $w_0 \in W_{Lx}$ that satisfies, for every $a_0, a_1 \in L \cup \{x\}$,

$$w_0 [a_0] \in C$$

$$w_0 [a_0] \,^\frown \varpi [a_1] \in {}^*C.$$

From this we also have, for every $a_0, a_1, b \in L \cup \{x\}$, that,

$$w_0 [a_0] \,^\frown \varpi [a_1] \,^{\frown *}\varpi [b] \in {}^{**}C.$$

Therefore, by transfer, there exists $w_1 \in W_{Lx}$ that satisfies, for every $a_0, a_1, a_2 \in L \cup \{x\}$:

$$w_0 [a_0] \in C$$

$$w_1 [a_1] \in C$$

$$w_0 [a_0] \,^\frown w_1 [a_1] \in C$$

$$w_0 [a_0] \,^\frown \varpi [a_2] \in {}^*C$$

$$w_1 [a_1] \,^\frown \varpi [a_2] \in {}^*C$$

$$w_0 [a_0] \,^\frown w_1 [a_1] \,^\frown \varpi [a_2] \in {}^*C.$$

Proceeding recursively, one can assume that at the $n$-th step elements $w_0, \ldots, w_{n-1}$ of $W_{Lx}$ have been defined such that, for every $n_1 < \cdots < n_k < n$ and $a_0, \ldots, a_{n-1}, a \in L \cup \{x\}$, one has that

$$w_{n_1} [a_{n_1}] \,^\frown \cdots \,^\frown w_{n_k} [a_{n_k}] \in C$$

$$w_{n_1} [a_{n_1}] \,^\frown \cdots \,^\frown w_{n_k} [a_{n_k}] \,^\frown \varpi [a] \in {}^*C.$$

From this one deduces also that for every $a, b \in L \cup \{x\}$ one has that

$$w_{n_1} [a_{n_1}] \,^\frown \cdots \,^\frown w_{n_k} [a_{n_k}] \,^\frown \varpi [a] \,^{\frown *}\varpi [b] \in {}^{**}C.$$

Hence, by transfer one obtains $w_n \in W_{Lx}$ such that for every $n_1 < \cdots < n_k \leq n$ and $a_0, \ldots, a_n, a \in L \cup \{x\}$, one has that

$$w_{n_1} [a_{n_1}] \,^\frown \cdots \,^\frown w_{n_k} [a_{n_k}] \in C$$

$$w_{n_1} [a_{n_1}] \,^\frown \cdots \,^\frown w_{n_k} [a_{n_k}] \,^\frown \varpi [a] \in {}^*C.$$

This concludes the recursive construction.

Theorem 7.6 now follows immediately from Proposition 7.10. Indeed, if $\{A_1, \ldots, A_r\}$ is a finite coloring of $W_L \cup W_{Lx}$, then there exist $1 \leq i, j \leq r$ such that $\upsilon \in {}^*A_i$ and $\varpi \in {}^*A_j$.


## Notes and References

Van der Waerden's theorem [130] is chronologically one of the first results in Ramsey theory, although preceded by the Hindman Cube Lemma [67] and by Schur's lemma on Schur triples [116]. Both van der Waerden's theorem and Schur's lemma were motivated by problems in modular arithmetic; see also [106, Chapter 2].

The Hales-Jewett theorem [66] is an abstract Ramsey-theoretic result motivated by the mathematical study of positional games such as "Tic-Tac-Toe" or "Go Moku". The original proof of Hales and Jewett from [66] was finitary and purely combinatorial. An infinitary proof was given by Bergelson et al. in [16]; see also [124, Chapter 2]. Combinatorial lines and combinatorial subspaces are also the object of the Graham–Rothschild theorem [59]. This was motivated by a conjecture of Rota on a geometric analogue of Ramsey's theorem. The conjecture was eventually established by Graham, Leeb, and Rothschild using similar methods [61].

# Chapter 8
# From Hindman to Gowers

The seminal results of Schur (Schur's Lemma) and Hilbert (Hilbert Cube Lemma) eventually led to the development of a whole research area at the interface between Ramsey theory and additive combinatorics. In this context, one studies which *additive* combinatorial configurations in $\mathbb{N}$ are *partition regular*, i.e. they can be found within a color of each finite coloring of $\mathbb{N}$. While van der Waerden's theorem on arithmetic progressions (discussed in the previous chapter) is the most famous early result in this area, several other additive configurations were later shown to be partition regular. Among these there are sets of finite sums of a finite sequence, which is the content of Folkman's theorem.

A conceptual leap was made in 1974 by Hindman when he established the *infinitary* version of Folkman's theorem. Hindman's theorem asserts that sets of finite sums of infinite sequences are partition regular, a result which is strictly stronger than its finitary counterpart. It had been observed by Galvin that such a statement is equivalent to the existence of an idempotent ultrafilter. Idempotent ultrafilters were not known to exist at the time and Hindman's original proof is purely combinatorial. It was later observed by Galvin that one can deduce the existence of idempotents ultrafilters from Ellis' theorem in topological dynamics, thus obtaining a short proof of Hindman's theorem.

Hindman's theorem was one of the first instances where ultrafilters and infinitary methods were shown to have a strong bearing on combinatorics in $\mathbb{N}$. Several refinements and generalizations were later obtained, almost all of which are rely on establishing the existence of certain ultrafilter configurations. Among these results is Gowers' Ramsey Theorem for $FIN_k$, which was motivated by a problem in the geometry of Banach spaces (namely oscillation-stability of $c_0$). In this chapter, we will present these fundamental results of Hindman and Gowers as well as the Milliken-Taylor Theorem, which is a simultaneous generalization of Ramsey's Theorem and Hindman's Theorem.

© Springer Nature Switzerland AG 2019
M. Di Nasso et al., *Nonstandard Methods in Ramsey Theory and Combinatorial Number Theory*, Lecture Notes in Mathematics 2239,
https://doi.org/10.1007/978-3-030-17956-4_8

## 8.1    Hindman's Theorem

Hindman's theorem is another fundamental pigeonhole principle, which considers the combinatorial configurations provided by sets of finite sums of infinite sequences.

**Definition 8.1**

1. Let $(c_n)$ be a sequence of elements of $\mathbb{N}$. Given a finite nonempty subset $F$ of $\mathbb{N}$, define $c_F := \sum_{n \in F} c_n$. Set $\mathrm{FS}((c_n)) := \{c_F \ : \ F \subseteq \mathbb{N} \text{ finite, nonempty}\}$. One can similarly define $c_F$ and $\mathrm{FS}(c_1, \ldots, c_k)$ for a finite sequence $(c_1, \ldots, c_k)$ of elements of $\mathbb{N}$.
2. We say that $A \subseteq \mathbb{N}$ is an *FS-set* if there is an infinite sequence $(c_n)$ of distinct elements from $\mathbb{N}$ such that $\mathrm{FS}((c_n)) \subseteq A$.

We begin this section by proving:

**Theorem 8.2 (Folkman's Theorem)**   *For any $m, r \in \mathbb{N}$, there is $n \in \mathbb{N}$ such that, for any $r$-coloring of $[1, n]$, there are $d_1, \ldots, d_m \in [1, n]$ such that $\mathrm{FS}(d_1, \ldots, d_m)$ is monochromatic.*

Folkman's theorem is a straightforward consequence of van der Waerden's theorem, as we show below following [117]. We say that $\mathrm{FS}(d_1, \ldots, d_m)$ is *weakly monochromatic* for some coloring if the color of $d_F$ depends only on $\max F$. Call the version of Folkman's theorem where monochromatic sets are replaced with weakly monochromatic sets the *weak version of Folkman's theorem*.

**Exercise 8.3**   The weak version of Folkman's theorem implies Folkman's theorem. (Hint: If there are $d_1, \ldots, d_m$ such that $\mathrm{FS}(d_1, \ldots, d_m)$ is weakly monochromatic for a coloring with $r$ colors, then there are $e_1, \ldots, e_n$, where $n = \lceil m/r \rceil$, such that $\mathrm{FS}(e_1, \ldots, e_n)$ is actually monochromatic for the coloring.)

*Proof (of Theorem 8.2)*   By Exercise 8.3, it suffices to prove the weak version of Folkman's theorem, which we prove by induction on $m$. The weak version of Folkman's theorem is clearly true for $m = 1$. Assume now that it is true for a given $m$ and we prove that it is true for $m + 1$. Fix $r$ and let $n$ witness the truth of the weak version of Folkman's theorem for $m$ and $r$, and let $p$ witness the truth of van der Waerden's theorem guaranteeing progressions of length $m + 1$ for colorings with $r$ colors. We claim that $p$ witnesses the truth of the weak version of Folkman's theorem for $m + 1$ and $r$. To see this, fix $p' \geq p$ and an $r$-coloring $c$ of $[1, p']$. Take a monochromatic subset $a + id, i = 0, 1, \ldots, m$, for the coloring $c$, say with color $q \in \{1, \ldots, r\}$. By the choice of $m$, there are $i_1, \ldots, i_n \in \{1, \ldots, m\}$ such that, setting $x_j := i_j d$, we have that $\mathrm{FS}(x_1, \ldots, x_d)$ is weakly monochromatic for $c$. Setting $x_{d+1} := a$, we see that $\mathrm{FS}(x_1, \ldots, x_d, x_{d+1})$ is also weakly monochromatic for $c$. Indeed, given $F \subseteq \{1, \ldots, d + 1\}$, either $d + 1 \notin F$, in which case $c(x_F)$ depends only on $\max F$ by assumption, or else $\max F = d + 1$, in which case $c(x_F) = q$.

An immediate consequence of Folkman's theorem is the statement that, for any finite coloring of $\mathbb{N}$, there are arbitrarily large finite sequences $(c_1, \ldots, c_n)$ in $\mathbb{N}$ such that $FS(c_1, \ldots, c_n)$ is monochromatic. The main result of this section, due to Hindman, allows us to find an *infinite* sequence $(c_n)$ in $\mathbb{N}$ such that $FS((c_n))$ is monochromatic. Just as the infinite Ramsey theorem cannot just be deduced from its finite form, Hindman's theorem cannot simply be deduced from Folkman's theorem.

**Theorem 8.4** *Suppose that $\alpha \in {}^*\mathbb{N}$ is $u$-idempotent. Then for every $A \subseteq \mathbb{N}$, if $\alpha \in {}^*A$, then $A$ is an FS-set.*

*Proof* We define by recursion $x_0 < x_1 < \cdots < x_n$ such that $x_F \in A$ and $x_F + \alpha \in {}^*A$ for any $F \subseteq \{0, 1, \ldots, n\}$. Note that, since $\alpha$ is idempotent, we also have that $x_F + \alpha + {}^*\alpha \in {}^{**}A$. Suppose that these have been defined up to $n$. The statement "there exists $w \in {}^*\mathbb{N}$ such that $w > x_n$ and, for every subset $F$ of $\{0, 1, 2, \ldots, n\}$, $x_F + w \in {}^*A$ and $x_F + w + {}^*\alpha \in {}^{**}A$" holds, as witnessed by $w = \alpha$. So, by transfer there exists $x_{n+1} \in \mathbb{N}$ larger than $x_n$ such that $x_F + x_{n+1} \in A$ and $x_F + x_{n+1} + \alpha \in {}^*A$ for any $F \subseteq \{0, 1, \ldots, n\}$. This concludes the recursive construction. $\qquad \blacksquare$

**Corollary 8.5 (Hindman)** *Any finite coloring of $\mathbb{N}$ has a monochromatic FS-set.*

*Proof* Let $\mathbb{N} := C_1 \sqcup \cdots \sqcup C_r$ be a finite coloring of $\mathbb{N}$. Let $\alpha$ be a $u$-idempotent element of ${}^*\mathbb{N}$ and let $i$ be such that $\alpha \in {}^*C_i$. The result now follows from the previous theorem. $\qquad \blacksquare$

**Lemma 8.6** *Suppose that $(c_n)$ is a sequence of distinct elements from $\mathbb{N}$. Then there is an idempotent $\alpha \in {}^*\mathbb{N}$ such that $\alpha \in {}^* FS((c_n))$.*

*Proof* For each $m$, let $U_m$ be the closed subset ${}^* FS((c_n)_{n \geq m})$ of ${}^*\mathbb{N}$. We have that $S := \bigcap_m U_m$ is a closed subset of ${}^*\mathbb{N}$ which is nonempty by compactness of ${}^*\mathbb{N}$ with the $u$-topology. We claim that $S$ is a $u$-subsemigroup of ${}^*\mathbb{N}$. Indeed, suppose that $\alpha, \beta \in S$ and let $\gamma \in {}^*\mathbb{N}$ such that $\alpha + {}^*\beta \sim \gamma$. We claim that $\gamma \in S$. Fix $m \in \mathbb{N}$. We must show that $\gamma \in {}^* FS((c_n)_{n \geq m})$ or, equivalently, $\alpha + {}^*\beta \in {}^{**} FS((c_n)_{n \geq m})$. Write $\alpha = c_F$ for some hyperfinite $F \subseteq \{n \in {}^*\mathbb{N} : n \geq m\}$. By transferring the fact that $\beta \in \bigcap_m S_m$, there is hyperfinite $G \subseteq \{n \in {}^{**}\mathbb{N} : n > \max(F)\}$ such that ${}^*\beta = c_G$, and so $\alpha + {}^*\beta = c_F + c_G \in {}^{**} FS((c_n)_{n \geq m})$.

It follows that $S$ is a nonempty closed $u$-subsemigroup of ${}^*\mathbb{N}$, whence, by Corollary 4.13, there is an idempotent $\alpha \in S$, which, in particular, implies that $\alpha \in {}^* FS((c_n))$. $\qquad \blacksquare$

**Corollary 8.7 (Strong Hindman's Theorem)** *Suppose that $C$ is an FS-set and $C$ is partitioned into finitely many pieces $C_1, \ldots, C_n$. Then some $C_i$ is an FS-set.*

*Proof* Take $(c_n)$ such that $FS((c_n)) \subseteq C$. Take $\alpha \in {}^*\mathbb{N}$ $u$-idempotent such that $\alpha \in {}^* FS((c_n))$. Then $\alpha \in {}^*C$ as well, whence $\alpha \in {}^*C_i$ for some $i = 1, \ldots, n$, and this $C_i$ is itself thus an FS-set. $\qquad \blacksquare$

**Exercise 8.8** Let Idem $:= \{\alpha \in {}^*\mathbb{N} : \alpha$ is $u$-idempotent$\}$. Prove that $\alpha \in \overline{\text{Idem}}$ if and only if: for every $A \subseteq \mathbb{N}$, if $\alpha \in {}^*A$, then $A$ is an FS-set. Here, $\overline{\text{Idem}}$ denotes the closure of Idem in the $u$-topology.

## 8.2   The Milliken-Taylor Theorem

We denote by $\mathbb{N}^{[m]}$ the set of subsets of $\mathbb{N}$ of size $m$. We identify $\mathbb{N}^{[m]}$ with the set of ordered $m$-tuples of elements of $\mathbb{N}$ increasingly ordered. If $F$, $G$ are finite subsets of $\mathbb{N}$, we write $F < G$ if either one of them is empty, or they are both nonempty and the maximum of $F$ is smaller than the minimum of $G$. Recall that for $F \subseteq \mathbb{N}$ finite, we use the notation $x_F$ for $\sum_{i \in F} x_i$, where we declare $x_F = 0$ when $F$ is empty.

The goal of this section is to prove the following:

**Theorem 8.9 (Milliken-Taylor)** *For any $m \in \mathbb{N}$ and finite coloring of $\mathbb{N}^{[m]}$, there exists an increasing sequence $(x_n)$ in $\mathbb{N}$ such that the set of elements of the form $\{x_{F_1}, \ldots, x_{F_m}\}$ for finite nonempty subsets $F_1 < \cdots < F_m$ of $\mathbb{N}$ is monochromatic.*

We note that the Milliken-Taylor theorem is a simultaneous generalization of Ramsey's theorem (by taking the finite sets $F_1, \ldots, F_m$ to have cardinality one) and Hindman's theorem (by taking $m = 1$).

The heart of the nonstandard approach is the following:

**Proposition 8.10** *Suppose that $m \in \mathbb{N}$ and $\alpha \in {}^*\mathbb{N}$ is $u$-idempotent. If $A \subset \mathbb{N}^{[m]}$ is such that $\{\alpha, {}^*\alpha, \ldots, {}^{*(m-1)}\alpha\} \in {}^{*m}A$, then there exists an increasing sequence $(x_n)$ in $\mathbb{N}$ such that $\{x_{F_1}, x_{F_2}, \ldots, x_{F_m}\} \in A$ for any finite nonempty subsets $F_1 < \cdots < F_m$ of $\mathbb{N}$.*

*Proof* We define by recursion an increasing sequence $(x_n)$ such that

$$\left\{ x_{F_1}, x_{F_2}, \ldots, x_{F_j}, \alpha, {}^*\alpha, \ldots, {}^{*(m-j-1)}\alpha \right\} \in {}^{*(m-j)}A$$

and

$$\left\{ x_{F_1}, x_{F_2}, \ldots, x_{F_{j-1}}, x_{F_j} + \alpha, {}^*\alpha, {}^{**}\alpha, \ldots, {}^{*(m-j)}\alpha \right\} \in {}^{*(m-j+1)}A$$

for every $1 \leq j \leq m$ and finite $F_1 < \cdots < F_j$ such that $F_1, \ldots, F_{j-1}$ are nonempty. It is clear that the sequence $(x_n)$ satisfies the conclusion of the proposition.

Suppose that we have constructed $x_1 < \cdots < x_{n-1}$ satisfying the recursive construction (where of course now $F_1, \ldots, F_j$ are subsets of $\{1, \ldots, n-1\}$). Since $\alpha$ is $u$-idempotent, we also have, for any $1 \leq j \leq m$ and $F_1, \ldots, F_j$ as above, that

$$\left\{ x_{F_1}, x_{F_2}, \ldots, x_{F_{j-1}}, x_{F_j} + \alpha + {}^*\alpha, {}^{**}\alpha, \ldots, {}^{*(m-j+1)}\alpha \right\} \in {}^{*(m-j+2)}A.$$

Therefore, by transfer there exists $x_n > x_{n-1}$ such that

$$\left\{ x_{F_1}, x_{F_2}, \ldots, x_{F_{j-1}}, x_{F_j} + x_n, \alpha, {}^*\alpha, \ldots, {}^{*(m-j-1)}\alpha \right\} \in {}^{*(m-j)}A$$

and

$$\left\{x_{F_1}, x_{F_2}, \ldots, x_{F_{j-1}}, x_{F_j} + x_n + \alpha, {}^*\alpha, \ldots, {}^{*(m-j)}\alpha\right\} \in {}^{*(m-j+1)}A$$

for any $1 \leq j \leq m$ and $F_1 < \cdots < F_j$ contained in $\{1, 2, \ldots, n-1\}$ such that $F_1, \ldots, F_{j-1}$ are nonempty. This concludes the recursive construction and the proof of the proposition.

Theorem 8.9 follows immediately from Proposition 8.10. Indeed, suppose $\mathbb{N}^{[m]} = A_1 \sqcup \cdots \sqcup A_r$ is a partition of $\mathbb{N}^{[m]}$. Fix $\alpha \in {}^*\mathbb{N}$ a $u$-idempotent. Let $i \in \{1, \ldots, r\}$ be such that $\left\{\alpha, {}^*\alpha, \ldots, {}^{*m-1}\alpha\right\} \in {}^{*m}A_i$. Then $A_i$ is the desired color.

Observe now that if $\lambda \in \mathbb{N}$ and $\alpha \sim \alpha + {}^*\alpha$, then $\lambda\alpha \sim \lambda\alpha + \lambda{}^*\alpha$. Hence the same proofs as above shows the following slight strengthening of Proposition 8.10, and hence of the Milliken-Taylor theorem.

**Proposition 8.11** *Suppose that $m \in \mathbb{N}$, $\lambda_1, \ldots, \lambda_m \in \mathbb{N}$, and $\alpha \in {}^*\mathbb{N}$ is $u$-idempotent. If $A \subset \mathbb{N}^{[m]}$ is such that $\left\{\alpha, {}^*\alpha, \ldots, {}^{*(m-1)}\alpha\right\} \in {}^{*m}A$, then there exists an increasing sequence $(x_n)$ in $\mathbb{N}$ such that $\left\{\lambda_1 x_{F_1}, \ldots, \lambda_m x_{F_m}\right\} \in A$ for any finite nonempty subsets $F_1 < \cdots < F_m$ of $\mathbb{N}$.*

**Theorem 8.12** *For any $m \in \mathbb{N}$, $\lambda_1, \ldots, \lambda_m \in \mathbb{N}$, and finite coloring of $\mathbb{N}^{[m]}$, there exists an increasing sequence $(x_n)$ in $\mathbb{N}$ such that the set of elements of the form $\left\{\lambda_1 x_{F_1}, \ldots, \lambda_m x_{F_m}\right\}$ for finite nonempty subsets $F_1 < \cdots < F_m$ of $\mathbb{N}$ is monochromatic.*

From the previous theorem, it is straightforward to deduce an "additive" version:

**Corollary 8.13** *For any $m \in \mathbb{N}$, $c_1, \ldots, c_m \in \mathbb{N}$, and finite coloring of $\mathbb{N}$, there exists an increasing sequence $(x_n)$ in $\mathbb{N}$ such that the set of elements of the form $c_1 x_{F_1} + \cdots + c_m x_{F_m}$ for finite nonempty subsets $F_1 < \cdots < F_m$ of $\mathbb{N}$ is monochromatic.*

## 8.3 Gowers' Theorem

The following partial semigroup was defined in Chap. 1. We recall the definition for the convenience of the reader.

**Definition 8.14** For $k \in \mathbb{N}$, we let $\mathrm{FIN}_k$ denote the set of functions $b : \mathbb{N} \to \{0, 1, \ldots, k\}$ with $\mathrm{Supp}(b)$ finite and such that $k$ belongs to the range of $b$. Here, $\mathrm{Supp}(b) := \{n \in \mathbb{N} : b(n) \neq 0\}$ is the *support* of $b$. We extend the definition of $\mathrm{FIN}_k$ to $k = 0$ by setting $\mathrm{FIN}_0$ to consist of the function on $\mathbb{N}$ that is identically 0.

Note that, after identifying a subset of $\mathbb{N}$ with its characteristic function, $\mathrm{FIN}_1$ is simply the set of nonempty finite subsets of $\mathbb{N}$. We endow $\mathrm{FIN}_k$ with a partial semigroup operation $(b_0, b_1) \mapsto b_0 + b_1$ which is defined only when Supp $(b_0) <$ Supp $(b_1)$.

By transfer, $*\mathrm{FIN}_k$ is the set of internal functions $b : *\mathbb{N} \to \{0, 1, \ldots, k\}$ with hyperfinite support that have $k$ in their range. The partial semigroup operation on $\mathrm{FIN}_k$ extends also to $*\mathrm{FIN}_k$. We say that $\alpha \in *\mathrm{FIN}_k$ is *cofinite* if its support is disjoint from $\mathbb{N}$. (This is a particular instance of Definition 4.14 in the case of the partial semigroup $\mathrm{FIN}_k$.) Thus, if $\alpha, \beta \in *\mathrm{FIN}_k$ are cofinite and $i < j$, then the sum $*^i\alpha + *^j\beta$ exists.

Gowers' original theorem considers the *tetris operation* $T : \mathrm{FIN}_k \to \mathrm{FIN}_{k-1}$ given by $T(b)(n) := \max \{b(n) - 1, 0\}$. In this section, we prove a more general version of Gowers' theorem by considering a wider variety of functions $\mathrm{FIN}_k \to \mathrm{FIN}_j$ for $j \le k$. First, for $k \in \mathbb{N}$, by a *regressive map on $k$* or *generalized tetris operation*, we mean a nondecreasing surjection $f : [0, k] \to [0, f(k)]$. Given a regressive map $f$ on $k$, one can define a corresponding operation $f : \mathrm{FIN}_k \to \mathrm{FIN}_{f(k)}$ by setting $f(b) := f \circ b$. Note also that if $l \le k$, then $f|_{[0,l]}$ is a regressive map on $l$, whence we can also consider $f : \mathrm{FIN}_l \to \mathrm{FIN}_{f(l)}$.

Given $n \in \mathbb{N}$, we set $\mathrm{FIN}_{[0,n]} := \bigcup_{k=0}^{n} \mathrm{FIN}_k$. Note that $\mathrm{FIN}_{[0,n]}$ is also a partial semigroup given by pointwise addition and defined on pairs of functions with disjoint supports. If $f$ is a regressive map on $n$, then as we already recalled, $f|_{[0,k]}$ is a regressive map on $k$ for $1 \le k \le n$, whence $f$ yields a function $f : \mathrm{FIN}_{[0,n]} \to \mathrm{FIN}_{[0,f(n)]}$.

Given a regressive map $f$ on $n$, we get the nonstandard extension $f : *\mathrm{FIN}_n \to *\mathrm{FIN}_{f(n)}$ and $f : *\mathrm{FIN}_{[0,n]} \to *\mathrm{FIN}_{[0,f(n)]}$. In addition, if $\alpha, \beta \in *\mathrm{FIN}_{[0,n]}$ are cofinite and $i < j$, then $*^i\alpha + *^j\beta$ exists and $f(*^i\alpha + *^j\beta) = f(*^i\alpha) + f(*^j\beta)$.

If $\alpha_k \in *\mathrm{FIN}_k$ for $k = 1, \ldots, n$, we say that a tuple $\langle \alpha_1, \ldots, \alpha_n \rangle$ is *coherent* if $f(\alpha_k) \sim \alpha_{f(k)}$ for all $k = 1, \ldots, n$ and all regressive maps $f$ on $n$. It is easy to verify that the set $Z$ of all cofinite coherent tuples is a compact $u$-semigroup. We note that $Z$ is nonempty. Indeed, let $\alpha_1 \in *\mathrm{FIN}_1$ be any cofinite element. For $k = 2, \ldots, n$, let $\alpha_k \in *\mathrm{FIN}_k$ have the same support as $\alpha_1$ and take only the values $0$ and $k$. It is immediate that $(\alpha_1, \ldots, \alpha_n) \in Z$.

Finally, we introduce some convenient notation. Given $\alpha_0, \alpha_1, \ldots, \alpha_j \in *\mathrm{FIN}_{[0,n]}$ and $j \in \mathbb{N}$, we set

$$\bigoplus_{i=0}^{j} \alpha_i := \alpha_0 + *\alpha_1 + \cdots + *^j\alpha_j.$$

Thus, if each $\alpha_i$ is cofinite and $f$ is a regressive map on $n$, we have the convenient equation

$$f\left(\bigoplus_{i=1}^{j} \alpha_i\right) = \bigoplus_{i=1}^{j} f(\alpha_i).$$

**Lemma 8.15** *Fix* $n \in \mathbb{N}$. *Then, for* $k = 1, \ldots, n$, *there exist cofinite* $u$-*idempotents* $\alpha_k \in {}^*\mathrm{FIN}_k$ *such that:*

1. $\langle \alpha_1, \ldots, \alpha_n \rangle$ *is a coherent tuple, and*
2. $\alpha_j + {}^*\alpha_k \sim \alpha_k + {}^*\alpha_j \sim \alpha_k$ *for every* $1 \leq j \leq k \leq n$.

*Proof* We define, by recursion on $k = 1, 2, \ldots, n$, a sequence of $u$-idempotents

$$\boldsymbol{\alpha}^{(k)} = (\alpha_1^{(k)}, \ldots, \alpha_n^{(k)}) \in Z$$

such that, for $1 \leq i \leq j \leq k \leq n$, one has that

(a) $\alpha_i^{(k)} \sim \alpha_i^{(j)}$,
(b) $\alpha_j^{(k)} + {}^*\alpha_i^{(k)} \sim \alpha_j^{(k)}$.

To begin the construction, let $\boldsymbol{\alpha}^{(1)}$ be any idempotent element of $Z$. Now suppose now that $k < n$ and $\boldsymbol{\alpha}^{(1)}, \ldots, \boldsymbol{\alpha}^{(k)}$ have been constructed satisfying (a) and (b). Consider the closed $u$-semigroup $Z_k$ consisting of sequences $\boldsymbol{\beta} = (\beta_1, \ldots, \beta_k) \in Z$ such that:

(i) $\beta_j \sim \alpha_j^{(k)}$ for $1 \leq j \leq k$, and
(ii) $\beta_j + {}^*\beta_i \sim \beta_j$ for $1 \leq i < j \leq n$ and $1 \leq i \leq k$.

We claim that $Z_k$ is nonempty. Indeed, we claim it contains the sequence $\boldsymbol{\beta} = (\beta_1, \ldots, \beta_k)$, where $\beta_j \in {}^*\mathrm{FIN}_j$ is such that

$$\beta_j \sim \bigoplus_{i=0}^{j-1} \alpha_{j-i}^{(k)}.$$

To see that $\boldsymbol{\beta}$ is coherent, fix a regressive map $f$ on $n$. For a given $j \in [1, k]$, we have that

$$f(\beta_j) \sim \bigoplus_{i=0}^{j-1} f(\alpha_{j-i}^{(k)}) \sim \bigoplus_{i=0}^{j-1} \alpha_{f(j-i)}^{(k)} \sim \bigoplus_{i=0}^{f(j)-1} \alpha_{f(j)-i}^{(k)} \sim \beta_{f(j)}.$$

The second equivalence uses that $\boldsymbol{\alpha}^{(k)}$ is coherent, while the third equivalence uses that $f$ is a regressive map and that $\boldsymbol{\alpha}^{(k)}$ is a $u$-idempotent. Next observe that, since $\boldsymbol{\alpha}^{(k)}$ satisfies (b), we have that $\beta_j \sim \bigoplus_{i=0}^{j-k} \alpha_{j-i}^{(k)}$ for $j = 1, \ldots, n$, and, moreover, that $\beta_j \sim \alpha_j^{(k)}$ for $j = 1, 2, \ldots, k$. Thus, if $1 \leq i < j \leq n$ and $1 \leq i \leq k$, it follows that

$$\beta_j + {}^*\beta_i \sim \bigoplus_{i=0}^{j-k} \alpha_{j-i}^{(k)} + {}^{*k}\alpha_i \sim \bigoplus_{i=0}^{j-k} \alpha_{j-i}^{(k)},$$

where the last equivalence follows from (b). This concludes the proof that $\beta$ belongs to $Z_k$.

Since $Z_k$ is a nonempty closed $u$-semigroup, it contains an idempotent $\boldsymbol{\alpha}^{(k+1)}$. It is clear that $\boldsymbol{\alpha}^{(k+1)}$ satisfies (a) and (b). This concludes the recursive construction.

For $k = 1, \ldots, n$, we fix $\alpha_k \in {}^*\mathrm{FIN}_k$ such that

$$\alpha_k \sim \bigoplus_{i=1}^{k} \alpha_i^{(i)}.$$

We claim that $\alpha_1, \ldots, \alpha_n$ are as in the conclusion of the lemma. Towards this end, first fix a regressive map $f$ on $n$. We then have that

$$f\left(\alpha_j\right) \sim \bigoplus_{i=1}^{k} f(\alpha_i^{(i)}) \sim \bigoplus_{i=1}^{k} \alpha_{f(i)}^{(i)} \sim \bigoplus_{i=1}^{f(k)} \alpha_i^{(i)} \sim \alpha_{f(j)},$$

where the second to last step uses the fact that $f$ is a regressive map, that the $\alpha_i^{(k)}$'s are $u$-idempotent, and that (a) holds. We thus have that $\alpha_1, \ldots, \alpha_n$ are coherent. We now show that (2) holds. Fix $1 \leq j \leq k \leq n$. We then have

$$\alpha_k + {}^*\alpha_j \sim \bigoplus_{i=1}^{k} \alpha_i^{(i)} + \bigoplus_{i=1}^{j} \alpha_i^{(i)} \sim \bigoplus_{i=1}^{k} \alpha_i^{(i)} \sim \alpha_k,$$

where the second to last equivalence repeatedly uses the fact that $\alpha_k^{(k)} + {}^*\alpha_i^{(i)} \sim \alpha_k^{(k)}$ for $1 \leq i \leq k$. A similar computation shows that $\alpha_j + {}^*\alpha_k \sim \alpha_k$, establishing (2) and finishing the proof of the lemma.

We say that a sequence $(x_i)$ in $\mathrm{FIN}_n$ is a *block sequence* if $\mathrm{Supp}\,(x_i) < \mathrm{Supp}\,(x_j)$ for $i < j$.

**Theorem 8.16** *Suppose that $\alpha_k \in {}^*\mathrm{FIN}_k$ for $k = 1, 2, \ldots, n$ are as in the previous lemma. For $k = 1, \ldots, n$, suppose that $A_k \subset \mathrm{FIN}_k$ is such that $\alpha_k \in {}^*A_k$. Then there exists a block sequence $(x_i)$ in $\mathrm{FIN}_n$ such that, for every finite sequence $f_1, \ldots, f_\ell$ of regressive maps on $n$, we have $f_1\,(x_1) + \cdots + f_\ell\,(x_\ell) \in A_{\max(f_1(n), \ldots, f_\ell(n))}$.*

*Proof* By recursion on $d$, we define a block sequence $(x_d)$ in $\mathrm{FIN}_n$ such that, for every sequence $f_1, \ldots, f_{d+1}$ of regressive maps $n$, we have

$$f_1\,(x_1) + \cdots + f_d\,(x_d) \in A_{\max(f_1(n), \ldots, f_d(n))}$$

and

$$f_1\,(x_1) + \cdots + f_d\,(x_d) + f_{d+1}\,(\alpha_n) \in {}^*A_{\max(f_1(n), \ldots, f_{d+1}(n))}.$$

Suppose that $x_1, \ldots, x_d$ has been constructed satisfying the displayed properties. Suppose that $f_1, \ldots, f_{d+2}$ are regressive maps on $n$. Then since

$$f_{d+1}(\alpha_n) + f_{d+2}({}^*\alpha_n) \sim \alpha_{f_{d+1}(n)} + {}^*\alpha_{f_{d+2}(n)} \sim \alpha_{\max(f_{d+1}(n), f_{d+2}(n))} \sim f_{d+p}(\alpha_n),$$

where $p \in \{1, 2\}$ is such that $\max(f_{d+1}(n), f_{d+2}(n)) = f_{d+p}(n)$, the inductive hypothesis allows us to conclude that

$$f_1(x_1) + \cdots + f_{d-1}(x_d) + f_{d+1}(\alpha_n) + f_{d+2}({}^*\alpha_n) \in {}^{**}A_{\max(f_1(n), \ldots, f_{d+2}(n))}.$$

Therefore, by transfer, we obtain $x_{d+1} \in \mathrm{FIN}_n$ such that $\mathrm{Supp}(x_{d+1}) > \mathrm{Supp}(x_d)$, and, for any sequence $f_1, \ldots, f_{d+2}$ of regressive maps on $n$, we have that

$$f_1(x_1) + \cdots + f_d(x_{d+1}) \in A_{\max(f_1(n), \ldots, f_{d+1}(n))}$$

and

$$f_1(x_1) + \cdots + f_{d+1}(x_{d+1}) + f_{d+2}(\alpha_n) \in {}^*A_{\max(f_1(n), \ldots, f_{d+2}(n))}.$$

This concludes the recursive construction.

**Corollary 8.17 (Generalized Gowers)** *For any finite coloring of* $\mathrm{FIN}_n$*, there exists a block sequence* $(x_l)$ *in* $\mathrm{FIN}_n$ *such that the set of elements of the form* $f_1(x_1) + \cdots + f_\ell(x_\ell)$ *for* $\ell \in \mathbb{N}$ *where* $f_1, \ldots, f_\ell$ *and regressive maps on* $n$ *such that* $n = \max(f_1(n), \ldots, f_\ell(n))$*, is monochromatic.*

*Proof* If $\mathrm{FIN}_n = B_1 \sqcup \cdots \sqcup B_r$ is a partition of $\mathrm{FIN}_n$, apply the previous theorem with $A_n := B_i$ where $\alpha_n \in {}^*B_i$.

Gowers' original theorem is a special case of the previous corollary by taking each $f_i$ to be an iterate of the tetris operation. One can also obtain a common generalization of Gowers' theorem and the Milliken-Taylor theorem. We let $\mathrm{FIN}_k^{[m]}$ be the set of $m$-tuples $(x_1, \ldots, x_m)$ in $\mathrm{FIN}_k$ such that $\mathrm{Supp}(x_i) < \mathrm{Supp}(x_j)$ for $1 \leq i < j \leq m$. Suppose that $(x_d)$ is a sequence in $\mathrm{FIN}_n$. Suppose that $F = \{a_1, \ldots, a_r\}$ is a finite nonempty subset of $\mathbb{N}$. We let $\mathscr{S}(F, k)$ be the set of tuples $f = (f_j)_{j \in F}$ such that $f_j : \{0, 1, \ldots, n\} \to \{0, 1, \ldots, k_j\}$ is a nondecreasing surjection and $\max\{k_j : j \in F\} = k$. For such an element $f$ we let $x_f$ be the sum $f_{a_1}(x_{a_1}) + \cdots + f_{a_r}(x_{a_r})$. When $F$ is empty, by convention we let $\mathscr{S}(F, k)$ contain a single element $f = \varnothing$, and in such case $x_f = 0$.

**Theorem 8.18** *Let* $\langle \alpha_1, \ldots, \alpha_n \rangle$ *be as in Lemma 8.15. Suppose that* $A_k \subset \mathrm{FIN}_k^{[m]}$ *for* $k = 1, 2, \ldots, n$ *is such that* $(\alpha_k, {}^*\alpha_k, \ldots, {}^{*(m-1)}\alpha_k) \in {}^{*m}A_k$*. Then there exists a block sequence* $(x_d)$ *in* $\mathrm{FIN}_n$ *such that, given* $k \in \{1, \ldots, n\}$*, nonempty finite subsets* $F_1 < \cdots < F_m$ *of* $\mathbb{N}$*, and* $f_i \in \mathscr{S}(F_i, k)$ *for* $i = 1, \ldots, m$*, we have that* $\{x_{f_1}, \ldots, x_{f_m}\} \in A_k$.

*Proof* We define by recursion a block sequence $(x_d)$ in $\mathrm{FIN}_n$ such that, for all $k \in \{1, \ldots, n\}$, all $1 \leq j \leq m$, all finite $F_1, \ldots, F_j \subseteq \mathbb{N}$ with $F_1 < \cdots < F_j$ and $F_1, \ldots, F_{j-1}$ nonempty, and all $f_i \in \mathscr{S}(F_i, k)$, we have

$$\left\{ x_{f_1}, x_{f_2}, \ldots, x_{f_j}, \alpha_k, {}^*\alpha_k, \ldots, {}^{*(m-j-1)}\alpha_k \right\} \in {}^{*(m-j)}A_k$$

and

$$\left\{ x_{f_1}, x_{f_2}, \ldots, x_{f_{j-1}}, x_{f_j} + \alpha_k, {}^*\alpha_k, {}^{**}\alpha_k, \ldots, {}^{*(m-j)}\alpha_k \right\} \in {}^{*(m-j+1)}A_k.$$

It is clear that the sequence $(x_d)$ is as desired.

Suppose that $x_1, \ldots, x_d$ have been constructed satisfying the above assumption. From the properties of the sequence $\alpha_1, \ldots, \alpha_n$, we see that the second condition also implies, for all $1 \leq s \leq k$:

$$\{ x_{f_1}, \ldots, x_{f_j} + \alpha_k + {}^*\alpha_s, {}^{**}\alpha_k, \ldots, {}^{*(m-j+1)}\alpha_k \} \in {}^{*(m-j+2)}A_k$$

and

$$\{ x_{f_1}, \ldots, x_{f_j} + \alpha_s + {}^*\alpha_k, {}^{**}\alpha_k, \ldots, {}^{**(m-j+1)}\alpha_k \} \in {}^{*(m-j+2)}A_k.$$

It follows from transfer that we can find $x_{d+1}$ with $\mathrm{Supp}(x_{d+1}) > \mathrm{Supp}(x_d)$ as desired.

## Notes and References

Hindman's theorem on finite sums, initially conjectured by Graham and Rothschild in [59], was first proved by Hindman by purely combinatorial methods [69]. It had been previously observed by Galvin—see also [68]—that the existence of an idempotent ultrafilter (which was unknown at the time) yields the conclusion of Hindman's theorem. The existence of idempotent ultrafilters was later established by Glazer; see [30]. Remarkably, Hindman's original combinatorial proof was significantly more technical and required a substantial amount of bookkeeping. Another short proof of Hindman's theorem was also obtained shortly later by Baumgartner [6].

Gowers' theorem [57] was motivated by a problem on the geometry of the Banach space $c_0$. While Gowers' original proof was infinitary and used ultrafilter methods, explicit purely combinatorial proofs of the corresponding finitary statement were later obtained by Ojeda-Aristizabal [105] and Tyros [127]. The more general version of Gowers' theorem presented in this chapter was established in [97]. This answered a question of Bartošová and Kwiatkowska from [5], where the corresponding finitary version is proved with different methods.

# Chapter 9
# Partition Regularity of Equations

Diophantine equations have been studied throughout the history of mathematics. Initially considered by Diophantus of Alexandria, a Hellenistic mathematician from the third century CE, they include the famous Fermat's Last Theorem, stated by Pierre de Fermat in the seventeenth century and eventually proved by Andrew Wiles in 1994. Hilbert's tenth problem asked to provide an algorithm or procedure to decide whether a given Diophantine equation admits integer solutions. Such an algorithm was shown not to exist by Yuri Matiyasevich in 1970, building on previous work of Martin Davis, Hilary Putnam, and Julia Robinson.

Deciding whether a given Diophantine equation admits integer solutions is in general a very hard problem, as made precise by Matiyasevich's theorem. Even less is known about the problem of deciding whether a Diophantine equation admits not just one, but "many" integer solutions, in the sense that any finite coloring admits a monochromatic solution (i.e. it is partition regular). Even the linear case of such a problem is nontrivial (it has van der Waerden's theorem as a particular case), and it was settled by Rado in the 1930s. Progress beyond the linear case has been scattered and only few cases are known to this day. Perhaps the most well-known open problem concerns the Pythagorean equation $x^2 + y^2 = z^2$. By means of a computer-assisted proof, it was recently shown that it is regular for two-partitions, but the general case remains so far unknown. It is also unknown whether there exists an algorithm to decide whether a given Diophantine equation is partition regular.

In this chapter, we investigate how nonstandard methods can be used to prove the partition regularity of some Diophantine equations as well as how they can be used to establish that some Diophantine equations are not partition regular.

© Springer Nature Switzerland AG 2019
M. Di Nasso et al., *Nonstandard Methods in Ramsey Theory and Combinatorial Number Theory*, Lecture Notes in Mathematics 2239, https://doi.org/10.1007/978-3-030-17956-4_9

## 9.1  Characterizations of Partition Regularity

Let $F(X_1, \ldots, X_n)$ be a function (here we will be interested in the case when $F$ is a polynomial over $\mathbb{Z}$). We begin with a proposition giving a nonstandard characterization of ultrafilters, all of whose members contain zeroes of $F$.

**Proposition 9.1** *Suppose that $\mathcal{U} \in \beta\mathbb{N}$. The following are equivalent:*

1. *For every $A \in \mathcal{U}$, there are [distinct] $x_1, \ldots, x_n \in A$ such that $F(x_1, \ldots, x_n) = 0$.*
2. *There exists $k \in \mathbb{N}$ and [distinct] $\alpha_1, \ldots, \alpha_n \in {}^{k*}\mathbb{N}$ such that $\mathcal{U} = \mathcal{U}_{\alpha_i}$ for all $i = 1, \ldots, n$ and $F(\alpha_1, \ldots, \alpha_n) = 0$.*

*Proof* First assume that (1) holds. For $A \in \mathcal{U}$, set

$$X_A := \left\{ (\alpha_1, \ldots, \alpha_n) \in {}^*\mathbb{N}^n \; : \; [\bigwedge_{i \neq j} \alpha_i \neq \alpha_j] \right.$$

$$\left. \wedge \bigwedge_i \alpha_i \in {}^*A \wedge F(\alpha_1, \ldots, \alpha_n) = 0 \right\}.$$

It is clear that the family $\{X_A \mid A \in \mathcal{U}\}$ has the finite intersection property, so by saturation there is $(\alpha_1, \ldots, \alpha_n) \in \bigcap_{A \in \mathcal{U}} X_A$; this tuple witnesses the truth of (2) with $k = 1$.

Conversely, suppose that (2) holds, and pick $\alpha_1, \ldots, \alpha_n \in {}^{k*}\mathbb{N}$ as given by the hypotheses. Suppose that $A \in \mathcal{U}$, and hence $\alpha_1, \ldots, \alpha_n \in {}^{k*}A$. Then the statement "there exist [distinct] $x_1, \ldots, x_n \in {}^{k*}A$ such that $F(x_1, \ldots, x_n) = 0$" holds, as witnessed by $\alpha_1, \ldots, \alpha_n$, and the desired conclusion follows by transfer. $\qquad\square$

**Definition 9.2** An ultrafilter $\mathcal{U}$ is a *witness* of the [injective] partition regularity of the equation $F(X_1, \ldots, X_n) = 0$ when $\mathcal{U}$ satisfies the equivalent conditions of Proposition 9.1. In this case, we also simply say that $\mathcal{U}$ is an [injective] *F-witness*. Similarly, if $\eta \in {}^*\mathbb{N}$ is such that $\mathcal{U}_\eta$ is an [injective] $F$-witness, then we also say that $\eta$ is an [injective] $F$-witness.

We now connect this notion with the standard Ramsey-theoretic notion of partition regular equation.

**Definition 9.3** A function $F(X_1, \ldots, X_n)$ is said to be *[injectively] partition regular* (on the natural numbers $\mathbb{N}$) if, for every finite partition $\mathbb{N} = C_1 \sqcup \cdots \sqcup C_r$, there exists $i \in \{1, \ldots, r\}$ and there exist [distinct] $x_1, \ldots, x_n \in C_i$ such that $F(x_1, \ldots, x_n) = 0$. In this case, we also say that the equation $F(x_1, \ldots, x_n) = 0$ is [injectively] partition regular.

**Proposition 9.4** *$F(X_1, \ldots, X_n)$ is [injectively] partition regular if and only if there is an [injective] $F$-witness if and only there exists $k \in \mathbb{N}$ and [distinct] $u$-equivalent $\alpha_1 \sim \ldots \sim \alpha_n$ in ${}^{k*}\mathbb{N}$ such that $F(\alpha_1, \ldots, \alpha_n) = 0$.*

*Proof* First suppose that $F(X_1, \ldots, X_n) = 0$ is [injectively] partition regular. Given $A \subseteq \mathbb{N}$, consider the set

$$Y_A := \{(\alpha_1, \ldots, \alpha_n) \in {}^*\mathbb{N}^n : [\bigwedge_{i \neq j} \alpha_i \neq \alpha_j]$$

$$\wedge \bigwedge_{i,j} (\alpha_i \in {}^*A \leftrightarrow \alpha_j \in {}^*A) \wedge F(\alpha_1, \ldots, \alpha_n) = 0\}.$$

Observe that the family $(Y_A)_{A \subseteq \mathbb{N}}$ has the finite intersection property. Indeed, given $A_1, \ldots, A_m \subseteq \mathbb{N}$, let $C_1, \ldots, C_k$ be the atoms of the Boolean algebra generated by $A_1, \ldots, A_m$. Since the equation $F(X_1, \ldots, X_n) = 0$ is [injectively] partition regular, there is $i \in \{1, \ldots, k\}$ and [distinct] $x_1, \ldots, x_n \in C_i$ such that $F(x_1, \ldots, x_n) = 0$; it follows that $(x_1, \ldots, x_n) \in \bigcap_{i=1}^m Y_{A_i}$. Thus, by saturation, there is $(\alpha_1, \ldots, \alpha_n) \in \bigcap_{A \subseteq \mathbb{N}} Y_A$. Then these $\alpha_1, \ldots, \alpha_n \in {}^*\mathbb{N}$ are $u$-equivalent and $F(\alpha_1, \ldots, \alpha_n) = 0$. Clearly, $\mathcal{U} = \mathcal{U}_{\alpha_i}$ is the desired $F$-witness.

The converse direction is trivial by the property of ultrafilter.

As an example, let us give a nonstandard proof of the following result, first shown by Brown and Rödl [22].

**Theorem 9.5** *A homogeneous equation $P(X_1, \ldots, X_n) = 0$ is [injectively] partition regular if and only if the corresponding equation with reciprocals $P(1/X_1, \ldots, 1/X_n) = 0$ is [injectively] partition regular.*

*Proof* Assume first that the homogeneous equation $P(X_1, \ldots, X_n) = 0$ is partition regular, and let $\alpha_1 \sim \ldots \sim \alpha_n$ be $u$-equivalent hypernatural numbers such that $P(\alpha_1, \ldots, \alpha_n) = 0$. Pick any infinite $\xi \in {}^*\mathbb{N}$. Note that ${}^*\xi > \alpha_i$, and so the factorial $({}^*\xi)! = {}^*(\xi!)$ is a multiple of $\alpha_i$ for all $i$. Let $\zeta_i := {}^*(\xi!)/\alpha_i \in {}^{**}\mathbb{N}$. Then $\alpha_1 \sim \ldots \sim \alpha_n \Rightarrow \zeta_1 \sim \ldots \sim \zeta_n$ and $P(1/\zeta_1, \ldots, 1/\zeta_n) = (1/{}^*(\xi!))^d P(\alpha_1, \ldots, \alpha_n) = 0$, where $d$ is the degree of $P$. This shows that $P(1/X_1, \ldots, 1/X_n)$ is partition regular. Clearly, by the same argument also the converse implication follows. Finally, note that the $\alpha_i$'s are distinct if and only if the $\zeta_i$'s are distinct, and so the equivalence holds also in the injective case.

## 9.2  Rado's Theorem

In this section we use the characterization of partition regularity shown above to prove the following version of the classical theorem of Rado for a single equation[1]:

---

[1] The general form of *Rado's Theorem* for a single equation states that $c_1 X_1 + \cdots + c_k X_k = 0$ is partition regular if and only there exists a nonempty set of indexes $I \subseteq \{1, \ldots, k\}$ with $\sum_{i \in I} c_i = 0$.

**Theorem 9.6** *Suppose that $k > 2$ and $c_1, \ldots, c_k \in \mathbb{Z}$ are such that $c_1 + \cdots + c_k = 0$. Then the equation $c_1 X_1 + \cdots + c_k X_k = 0$ is injectively partition regular.*

As we will see, in this setting $u$-idempotent elements will play a crucial role. Let us start with a nonstandard proof of a simple particular case, which was first proved in [15].

**Theorem 9.7** *Let $\mathscr{U}$ be any idempotent ultrafilter. Then $2\mathscr{U} \oplus \mathscr{U}$ is a witness of the injective partition regularity of the equation $X_1 - 2X_2 + X_3 = 0$.*[2]

*Proof* Pick a $u$-idempotent point $\xi \in {}^*\mathbb{N}$ such that $\mathscr{U}_\xi = \mathscr{U}$, and consider the following three distinct elements of ${}^{***}\mathbb{N}$: $\alpha_1 := 2\xi + {}^{**}\xi$, $\alpha_2 := 2\xi + {}^*\xi + {}^{**}\xi$, and $\alpha_3 := 2\xi + 2{}^*\xi + {}^{**}\xi$. Then $2\mathscr{U} \oplus \mathscr{U} = \mathscr{U}_{\alpha_1} = \mathscr{U}_{\alpha_2} = \mathscr{U}_{\alpha_3}$ and $\alpha_1 - 2\alpha_2 + \alpha_3 = 0$.

Let us now prove a strengthening of Rado's theorem below, grounded on idempotent ultrafilters. First, given a polynomial $P(X) := \sum_{j=0}^{n} b_j X^j \in \mathbb{Z}[X]$ and $\xi \in {}^*\mathbb{Z}$, set $\tilde{P}(\xi) := \sum_{j=0}^{n} b_j {}^{j*}\xi \in {}^{(j+1)*}\mathbb{Z}$. We note the following corollary of Proposition 9.4.

**Corollary 9.8** *Suppose that $c_1, \ldots, c_k \in \mathbb{Z}$ are such that there exist [distinct] polynomials $P_1(X), \ldots, P_k(X) \in \mathbb{Z}[X]$ and $\xi, \eta \in {}^*\mathbb{N}$ for which*

1. *$c_1 P_1(X) + \cdots + c_k P_k(X) = 0$, and*
2. *$\tilde{P}_i(\xi) \sim \eta$ for each $i = 1, \ldots, k$.*

*Then $\mathscr{U}_\eta$ witnesses that $c_1 X_1 + \cdots + c_k X_k = 0$ is [injectively] partition regular.*

*Proof* For each $i = 1, \ldots, k$, let $\alpha_i := \tilde{P}_i(\xi)$; by assumption, for each $i$ we have $\mathscr{U}_{\alpha_i} = \mathscr{U}_\eta$. It is also clear that $c_1 \alpha_1 + \cdots + c_k \alpha_k = 0$. By Proposition 9.4, we have that $\mathscr{U}_\eta$ witnesses the partition regularity of $c_1 X_1 + \cdots + c_k X_k = 0$.

Suppose in addition that the $P_i$'s are distinct; to conclude injective partition regularity, we must show that the $\alpha_i$'s are distinct. Suppose that $\alpha_i = \alpha_j$, that is, $\tilde{P}_i(\xi) = \tilde{P}_j(\xi)$. Write $P_i(X) := \sum_{l=0}^{m} r_l X^l$ and $P_j(X) := \sum_{l=0}^{m} s_l X^l$, where at least one between $r_m$ and $s_m$ is nonzero. We then have that $(r_m - s_m)^{m*}\xi = -\sum_{l=0}^{m-1}(r_l - s_l)^{l*}\xi$. The only way that this is possible is that $r_m = s_m$. Continuing inductively in this manner, we see that $P_i = P_j$, yielding the desired contradiction.

In light of the previous corollary, it will be useful to find a standard condition on a family of polynomials $P_1, \ldots, P_k \in \mathbb{Z}[X]$ such that, for every idempotent $\xi \in {}^*\mathbb{N}$, we have that all $\tilde{P}_i(\xi)$'s are $u$-equivalent. The next definition captures such a condition.

---

[2]This theorem is essentially the content of Exercise 4.11.

**Definition 9.9** Following [38], we define the equivalence relation $\approx_u$ on finite strings of integers to be the smallest equivalence relation satisfying the following three properties:

- $\emptyset \approx_u \langle 0 \rangle$;
- If $a \in \mathbb{Z}$, then $\langle a \rangle \approx_u \langle a, a \rangle$;
- If $\sigma \approx_u \sigma'$ and $\tau \approx_u \tau'$, then concatenations $\sigma ^\frown \tau \approx_u \sigma' ^\frown \tau'$.

If $P, Q \in \mathbb{Z}[X]$ are polynomials, then we write $P \approx_u Q$ to mean that their strings of coefficients are $u$-equivalent.

**Lemma 9.10** *Let $P, Q \in \mathbb{Z}[X]$ have positive leading coefficient. If $P \approx_u Q$, then for every idempotent $\xi \in {}^*\mathbb{N}$, we have $\tilde{P}(\xi) \sim \tilde{Q}(\xi)$.*

*Proof* Fix an idempotent $\xi \in {}^*\mathbb{N}$. The lemma follows from the following facts:

- $\sum_{j=0}^{m} a_j{}^{j*}\xi \sim \sum_{j=0}^{i} a_j{}^{j*}\xi + a_i{}^{(i+1)*}\xi + \sum_{j=i+1}^{m} a_j{}^{(j+1)*}\xi$;
- If $\sum_{j=0}^{m} a_j{}^{j*}\xi \sim \sum_{j=0}^{m'} a_j'{}^{j*}\xi$ and $\sum_{j=0}^{n} b_j{}^{j*}\xi \sim \sum_{j=0}^{n'} b_j'{}^{j*}\xi$, then

$$\sum_{j=0}^{m} a_j{}^{j*}\xi + \sum_{j=0}^{n} b_j{}^{(m+j+1)*}\xi \sim \sum_{j=0}^{m'} a_j'{}^{j*}\xi + \sum_{j=0}^{n'} b_j'{}^{(m'+j+1)*}\xi.$$

We should mention that the converse of the previous lemma is true in an even stronger form, namely that if $\tilde{P}(\xi) \sim \tilde{Q}(\xi)$ for *some* idempotent $\xi \in {}^*\mathbb{N}$, then $P \approx_u Q$. This follows from [100, Theorem T].

We can now give the nonstandard proof of the above mentioned version of Rado's theorem. In fact, we prove the more precise statement:

**Theorem 9.11** *Suppose that $k > 2$ and $c_1, \ldots, c_k \in \mathbb{Z}$ are such that $c_1 + \cdots + c_k = 0$. Then there exists $a_0, \ldots, a_{k-2} \in \mathbb{N}$ such that, for every idempotent ultrafilter $\mathscr{U}$, we have that $a_0 \mathscr{U} \oplus \cdots \oplus a_{k-2} \mathscr{U}$ witnesses the injective partition regularity of the equation $c_1 X_1 + \cdots + c_k X_k = 0$.*

*Proof* Without loss of generality, we will assume that $c_1 \geq c_2 \geq \cdots \geq c_k$. By Corollary 9.8 and Lemma 9.10, we need to find $a_0, \ldots, a_{k-2} \in \mathbb{N}$ and distinct $P_1(X), \ldots, P_k(X) \in \mathbb{Z}[X]$ such that $c_1 P_1(X) + \cdots + c_k P_k(X) = 0$ and such that $P_i(X) \approx_u \sum_{j=0}^{k-2} a_j X^j$ for each $i = 1, \ldots, k$. For appropriate $a_0, \ldots, a_{k-2}$, the following polynomials will be as needed:

- $P_1(X) := \sum_{j=0}^{k-2} a_j X^j + a_{k-2} X^{k-1}$;
- $P_i(X) := \sum_{j=0}^{k-i-1} a_j X^j + \sum_{j=k-i+1}^{k-1} a_{j-1} X^j$ for $2 \leq i \leq k-1$,
- $P_k(X) := a_0 + \sum_{j=1}^{k-1} a_{j-1} X^j$.

It is straightforward to check that $P_i(X) \approx_u \sum_{j=0}^{k-2} a_j X^j$ for each $i = 1, \ldots, k$. Furthermore, since $a_0, \ldots, a_{k-2}$ are nonzero, the polynomials $P_1(X), \ldots, P_k(X)$ are mutually distinct. It remains to show that there are $a_0, \ldots, a_{k-2} \in \mathbb{N}$ for which $c_1 P_1(X) + \cdots + c_k P_k(X) = 0$. Since $c_1 + \cdots + c_k = 0$, the constant and leading terms of $c_1 P_1(X) + \cdots + c_k P_k(X)$ are zero. So the equation $c_1 P_1(X) + \cdots + c_k P_k(X) = 0$ is equivalent to the system of equations $(c_1 + \cdots + c_{k-i}) \cdot a_{i-1} + (c_{k-i+1} + \cdots + c_k) \cdot a_{i-2}$ for $i = 2, 3, \ldots, k - 1$. One can then easily define recursively elements $a_0, a_1, \ldots, a_{k-2}$ satisfying all these equations.

## 9.3   Nonlinear Diophantine Equations: Some Examples

Although Rado's Theorem from the 1930s completely settled the partition regularity problem for linear Diophantine equations, progress on the nonlinear case has been scattered. However, it is worth stressing that in the last few years, many exciting new results have been proven, yielding promising research aimed towards general characterization theorems. (See the notes at the end of this chapter.) In this section, we only present a small selection of relevant nonlinear equations, and give nonstandard proofs about their partition regularity or non-partition regularity.

Let us start with following theorem of Hindman from 2011 [71].

**Theorem 9.12** *For any $m, n \in \mathbb{N}$, the equation $x_1 + \cdots + x_m = y_1 \cdots y_n$ is injectively partition regular.*

The idea of the nonstandard proof presented below is due to L. Luperi Baglini (see [94], where a generalization of that result is proved). The following proposition is the key idea.

**Proposition 9.13** *Suppose that $P(X_1, \ldots, X_n)$ is a homogeneous equation that is injectively partition regular. Then there is a multiplicatively idempotent $\mathcal{U} \in \beta\mathbb{N}$ (that is, $\mathcal{U} \odot \mathcal{U} = \mathcal{U}$) that witnesses the injective partition regularity of $P$.*

*Proof* Let $I_P$ be the set of $P$-witnesses. It suffices to show that $I_P$ is a nonempty, closed subsemigroup of $(\beta\mathbb{N}, \odot)$. $I_P$ is nonempty by definition. $I_P$ is closed since it consists of those ultrafilters whose members $A$ all satisfy the property of containing a tuple that is solution of $P$; see Exercise 1.13. Finally, we show that $I_P$ is closed under multiplication. In fact, we show that $I_P$ is a two-sided ideal. Suppose that $\mathcal{U} \in I_P$ and $\mathcal{V} \in \beta\mathbb{N}$. Take distinct $\alpha_1, \ldots, \alpha_n$ such that $\mathcal{U} = \mathcal{U}_{\alpha_i}$ for $i = 1, \ldots, n$ and $P(\alpha_1, \ldots, \alpha_n) = 0$. Also let $\beta$ be such that $\mathcal{V} = \mathcal{U}_\beta$. We then have that $\alpha_1 {}^*\beta, \ldots, \alpha_n {}^*\beta$ are distinct generators of $\mathcal{U} \odot \mathcal{V}$ and, setting $d$ to be the degree of $P$, we have

$$P(\alpha_1 {}^*\beta, \ldots, \alpha_n {}^*\beta) = {}^*\beta^d P(\alpha_1, \ldots, \alpha_n) = 0.$$

It follows that $\mathcal{U} \odot \mathcal{V}$ belongs to $I_P$. The proof that $I_P$ is a right-ideal is similar and left to the reader. The proof is concluded by applying Ellis' Theorem (Theorem 1.23) to the compact semitopological semigroup $(I_P, \odot)$.

We now prove Theorem 9.12 in the simple case $m = 2$ and $n = 3$. Since $x_1 + x_2 - y = 0$ is homogeneous and injectively partition regular by Rado's theorem,

Proposition 9.13 implies that we may find a multiplicative idempotent ultrafilter $\mathscr{U}$ that witnesses the injective partition regularity of such an equation. Take distinct $\alpha_1, \alpha_2, \beta \in {}^*\mathbb{N}$ all of which generate $\mathscr{U}$ and for which $\alpha_1 + \alpha_2 = \beta$. For $i = 1, 2$, set $\gamma_i := \alpha_i {}^* \beta {}^{**} \beta$. Note that $\gamma_1$ and $\gamma_2$ are also distinct generators of $\mathscr{U}$ and $\gamma_1 + \gamma_2 = \beta {}^* \beta {}^{**} \beta$. Since $\mathscr{U}$ is multiplicatively idempotent, we have that $\beta$ is multiplicatively $u$-idempotent, and hence $\beta {}^* \beta {}^{**} \beta \sim \beta$. It follows that $\gamma_1 \sim \gamma_2 \sim \beta \sim {}^*\beta \sim {}^{**}\beta$ are witness the injective partition regularity of $x_1 + x_2 - y_1 \cdot y_2 \cdot y_3 = 0$.

Nonstandard methods have also played a role in establishing the non-partition regularity of equations. We present here simple examples of this type of results.

**Theorem 9.14 ([40])** *Let $P(x_1, \ldots, x_h) := a_1 x_1^{n_1} + \cdots + a_h x_h^{n_h}$, with $n_1 < \cdots < n_h$, where each $a_i \in \mathbb{Z}$ is odd and $h$ is odd. Then $P(x_1, \ldots, x_h) = 0$ is not partition regular.*

*Proof* Suppose, towards a contradiction, that there are $u$-equivalent $\xi_1, \ldots, \xi_h \in {}^*\mathbb{N}$ such that $P(\xi_1, \ldots, \xi_h) = 0$. Let $f, g : \mathbb{N} \to \mathbb{N}_0$ be such that, for all $x \in \mathbb{N}$, we have $x = 2^{f(x)} g(x)$ with $g(x)$ odd. Then, for each $i, j = 1, \ldots, h$, we have $f(\xi_i) \sim f(\xi_j)$. Set $\nu_i := f(\xi_i)$ and $\zeta_i := g(\xi_i)$.

We next claim that, for distinct $i, j \in \{1, \ldots, h\}$, we have $n_i \nu_i \neq n_j \nu_j$. Indeed, if $n_i \nu_i = n_j \nu_j$, then $n_i \nu_i = n_j \nu_j \sim n_j \nu_i$, whence $n_i \nu_i = n_j \nu_i$ by Proposition 3.5 and hence $\nu_i = 0$. Since the $\nu_k$'s are all $u$-equivalent, it follows that $\nu_k = 0$ for each $k$, whence each $\xi_i$ is odd. But then since $h$ is odd, we have that $P(\xi_1, \ldots, \xi_h)$ is odd, contradicting that $P(\xi_1, \ldots, \xi_h) = 0$.

By the previous paragraph, we can let $i \in \{1, \ldots, k\}$ be the unique index for which $n_i \nu_i < n_j \nu_j$ for all $j = 1, \ldots, k$ other than $i$. By factoring out $2^{n_i \nu_i}$ from the equation $P(\xi_1, \ldots, \xi_h) = 0$, we obtain the contradiction

$$0 = a_i \zeta_i^{n_i} + \sum_{j \neq i} a_j 2^{n_j \nu_j - n_i \nu_i} \zeta_j^{n_j} \equiv 1 \mod 2.$$

From the previous theorem, we see that many "Fermat-like" equations are not partition regular:

**Corollary 9.15** *Suppose that $k, m, n$ are distinct positive natural numbers. Then the equation $x^m + y^n = z^k$ is not partition regular.*

In [40], the previous corollary is extended to allow $m$ and $n$ to be equal, in which case the equations are shown to be not partition regular (as long as, in the case when $m = n = k - 1$, one excludes the trivial solution $x = y = z = 2$). The methods are similar to the previous proof.

To further illustrate the methods, we conclude by treating two simple cases of non-partition regular equations. Let us start with a result that was first proven by Csikivari, Gyarmati, and Sarkozy in [31]. The nonstandard proof given below uses the same idea as in [63].

**Theorem 9.16** *If one excludes the trivial solution $x = y = z = 2$, then the equation $x + y = z^2$ is not partition regular.*

*Proof* Suppose, towards a contradiction, that $\alpha, \beta, \gamma \in {}^*\mathbb{N}\backslash\mathbb{N}$ are such that $\alpha \sim \beta \sim \gamma$ and $\alpha + \beta = \gamma^2$. Without loss of generality, assume $\alpha \geq \beta$. Let $f : \mathbb{N} \to \mathbb{N}_0$ be the function defined by $2^{f(n)} \leq n < 2^{f(n)+1}$ for every $n$, and set $a := f(\alpha)$. Notice that $a$ is infinite, as otherwise $\alpha$ would be finite and $\alpha \sim \beta \sim \gamma$ would imply $\alpha = \beta = \gamma = 2$, contrary to our hypothesis. Now observe that

$$2^a \leq \alpha < \alpha + \beta = \gamma^2 \leq 2\alpha < 2 \cdot 2^{a+1} \Rightarrow 2^{\frac{a}{2}} < \gamma < 2^{\frac{a}{2}+1}.$$

This shows that either $f(\gamma) = \lfloor \frac{a}{2} \rfloor$ or $f(\gamma) = \lfloor \frac{a}{2} \rfloor + 1$. Since $a = f(\alpha) \sim f(\gamma)$, we have either $a \sim \lfloor \frac{a}{2} \rfloor$ or $a \sim \lfloor \frac{a}{2} \rfloor + 1$. In both cases we reach a contradiction as we would either have $a = \lfloor \frac{a}{2} \rfloor$ (which cannot occur since $a$ is positive), or $a = \lfloor \frac{a}{2} \rfloor + 1$ (and hence $a = 1$ or $a = 2$; the former is impossible and the latter has been excluded by hypothesis).

**Theorem 9.17** *The equation $x^2 + y^2 = z$ is not partition regular.*

*Proof* Notice first that the given equation does not have constant solutions. Then suppose, towards a contradiction, that $\alpha, \beta, \gamma$ are infinite hypernatural numbers such that $\alpha \sim \beta \sim \gamma$ and $\alpha^2 + \beta^2 = \gamma$. Notice that $\alpha, \beta, \gamma$ are even numbers, since they cannot all be odd. Then we can write

$$\alpha = 2^a \alpha_1, \quad \beta = 2^b \beta_1, \quad \gamma = 2^c \gamma_1,$$

with positive $a \sim b \sim c$ and with $\alpha_1 \sim \beta_1 \sim \gamma_1$ odd.

*Case 1* $a < b$. We then have that $2^{2a}(\alpha_1^2 + 2^{2b-2a}\beta_1^2) = 2^c \gamma_1$. Since $\alpha_1^2 + 2^{2b-2a}\beta_1^2$ and $\gamma_1$ are odd, it follows that $2a = c \sim a$, whence $2a = a$ by Proposition 3.5 and hence $a = 0$, a contradiction. If $b > a$ the proof is entirely similar.

*Case 2* $a = b$. In this case we have the equality $2^{2a}(\alpha_1^2 + \beta_1^2) = 2^c \gamma_1$. Since $\alpha_1, \beta_1$ are odd, $\alpha_1^2 + \beta_1^2 \equiv 2 \mod 4$, and so $2^c \gamma_1 = 2^{2a+1}\alpha_2$ for a suitable odd number $\alpha_2$. But then $2a + 1 = c \sim a$, whence $2a + 1 = a$, and we again obtain a contradiction.

## Notes and References

Rado's theorem is one of the first general results in Ramsey theory [109], building on previous work of Hilbert and Rado's advisor Schur. In particular, Rado's Theorem (in its extended version about systems of equations) subsumes van der Waerden's Theorem on arithmetic progressions. Since then, only fragmented and isolated progress has been obtained in the study of partition regularity of more general (nonlinear) equations, starting from the 1990s. Let us quickly mention here some relevant contributions.

In 1991 [87], Lefmann studied homogeneous polynomials and proved that, for every $n \in \mathbb{N}$, equations $c_1 X_1^{1/n} + \cdots + c_k X_k^{1/n} = 0$ are partition regular if and only if "Rado's condition" on coefficients hold (that is, $\sum_{i\in I} a_i = 0$ for some nonempty $I \subseteq \{1, \ldots, k\}$). In 1996, grounding on a density result by Fürstenberg and Sarkozy, Bergelson [13] showed the partition regularity of all Diophantine equations $X - Y = P(Z)$ where the polynomial $P(Z) \in \mathbb{Z}[Z]$ has no constant term. In 2010, by using algebra in the space of ultrafilters $\beta\mathbb{N}$, Bergelson [14] proved the partition regularity of the equation $X_1 + X_2 = Y_1 Y_2$. Independently, Hindman [71] showed the partition regularity of all equations of the form $\sum_{i=1}^{k} X_i = \prod_{i=1}^{k} Y_i$.

In the last few years, research on the nonlinear case has made significant progress; in particular, nonstandard methods have proven quite effective. In 2014, Luperi Baglini [94] improved on the above mentioned result of Hindman, and, by using nonstandard methods, proved that for every $F_i \subseteq \{1, \ldots, m\}$, the equation $\sum_{i=1}^{k} c_i X_i (\prod_{j\in F_i} Y_j) = 0$ is partition regular whenever $\sum_{i\in I} c_i = 0$ for some nonempty $I \subseteq \{1, \ldots, m\}$. In [40], Di Nasso and Riggio used nonstandard analysis to characterize a large class of Fermat-like equations that are not partition regular, the simplest cases being $X^m + Y^n = Z^k$ where $k \notin \{n, m\}$. In their paper [50] of 2017, Frantzikinakis and Host showed, as consequences of their structural theorem for multiplicative functions, that the quadratic equations $16X^2 + 9Y^2 = Z^2$ and $X^2 - XY + Y^2 = Z^2$ are partially partition regular in the variables $X$ and $Y$.

The study of partition regularity of Diophantine equations can be seen as a particular instance of the more general problem of establishing the partition regularity of arbitrary configurations. One outstanding such problem, recently settled positively by Moreira using topological dynamics [101], was the problem of partition regularity of the configuration $\{a, a+b, ab\}$ in $\mathbb{N}$. (It is still unknown at the time of writing whether the configuration $\{a, b, a+b, ab\}$ is partition regular in $\mathbb{N}$.) In the same paper, Moreira also showed the partition regularity of the equations $c_1 X_1^2 + \ldots + c_k X_k^2 = Y$ where $c_1 + \ldots + c_k = 0$ (so, e.g. $X^2 + Y = Z^2$ is partition regular). On the other hand, those equations are not partition regular when "Rado's condition" fails, as shown in [39]. Interesting results about exponential configurations were recently proved by Sahasrabudhe [114, 115], starting from the partition regularity of $\{a, b, a^b\}$. In [28], Chow, Lindqvist, and Prendiville characterized the partition regularity of classes of generalised Pythagorean equations by means of Rado's condition.

A breakthrough was obtained in [39], where general necessary criteria for partition regularity of large classes of Diophantine equations are obtained using nonstandard methods and iterated hyperextensions, as well as sufficient criteria using algebra in the Stone-Čech compactification. One example is the following:

- *A Diophantine equation of the form* $c_1 X_1 + \ldots + c_k X_k = P(Y)$ *where* $P$ *is a nonlinear polynomial with no constant term is PR if and only if "Rado's condition" holds in the linear part, i.e.* $\sum_{i\in I} a_i = 0$ *for some nonempty* $I \subseteq \{1, \ldots, k\}$.

So, e.g. $X - 2Y = Z^n$ is not PR (while $X - Y = Z^n$ is); and $X + Y = P(Z)$ is not partition regular for any nonlinear $P(Z) \in \mathbb{Z}[X]$ (while if $P(Z)$ takes even values on some integer, then in every finite coloring one has a solution where $X$ and $Y$ are monochromatic, as shown by Khalfalah and Szemerédi [85] in 2006); and equations

$X_1 - 2X_2 + X_3 = Y^n$ are PR, so that in every finite coloring of the natural numbers one finds monochromatic configurations of the form $\{a, b, c, 2a - b + c^n\}$; and so forth.

One of the most famous open problems in the area concerns the partition regularity of the Pythagorean equation $X^2 + Y^2 = Z^2$. In 2016, by means of a computer-assisted proof (dubbed "the longest mathematical proof ever" [86]), Heule, Kullmann and Marek proved that any 2-coloring of $\{1, 2, \ldots, 7825\}$ contains a monochromatic Pythagorean triple. However, the general problem with arbitrary finite partitions remains unsolved at the time of this writing.

# Part III
# Combinatorial Number Theory

# Chapter 10
# Densities and Structural Properties

Ramsey theory initially focused on the study of combinatorial properties that are "abundant" in the sense that can always be found in one of the pieces of any finite partition of a given structure. Such a notion can be strengthened by considering combinatorial configuration that can be found in any set that is "large" in a more generous quantitative sense. Typically, largeness is expressed in terms of natural notions of density, which are obtained via a limiting process from the relative density within finite portions of the structure. In the case of $\mathbb{N}$, the most natural choices are the *upper density* and the *Banach density*, which naturally arise in the study of dynamical systems and combinatorial number theory. In this chapter, we introduce such densities as well as some structural notions of largeness for sets of natural numbers. Along the way, we prove their nonstandard reformulations and develop some of the basic properties of these notions needed in later chapters. The chapter concludes with the nonstandard perspective on the Furstenberg correspondence.

## 10.1 Densities

In this section, $A$ and $B$ denote subsets of $\mathbb{N}$. Recall that $\delta(A, n) = \frac{|A \cap [1,n]|}{n}$.

**Definition 10.1**

1. The *upper density* of $A$ is defined to be

$$\overline{d}(A) := \limsup_{n \to \infty} \delta(A, n).$$

2. The *lower density* of $A$ is defined to be

$$\underline{d}(A) := \liminf_{n \to \infty} \delta(A, n).$$

© Springer Nature Switzerland AG 2019
M. Di Nasso et al., *Nonstandard Methods in Ramsey Theory
and Combinatorial Number Theory*, Lecture Notes in Mathematics 2239,
https://doi.org/10.1007/978-3-030-17956-4_10

3. If $\overline{d}(A) = \underline{d}(A)$, then we call this common value the *density of* $A$ and denote it by $d(A)$.

The following exercise concerns the nonstandard characterizations of the aforementioned densities.

**Exercise 10.2** Prove that

$$\overline{d}(A) = \max\{\mathrm{st}(\delta(A, N)) \,:\, N \in {}^*\mathbb{N}\backslash\mathbb{N}\} = \max\{\mu_N({}^*A) \,:\, N \in {}^*\mathbb{N}\backslash\mathbb{N}\},$$

where $\mu_N$ is the Loeb measure on $[1, N]$. State and prove the corresponding statement for lower density.

The previous exercise illustrates why the nonstandard approach to densities is so powerful. Indeed, while densities often "feel" like measures, they lack some of the key properties that measures possess. However, the nonstandard approach allows us to treat densities as measures, thus making it possible to use techniques from measure theory and ergodic theory.

There is something artificial in the definitions of upper and lower density in that one is always required to take samples from initial segments of the natural numbers. We would like to consider a more uniform notion of density which allows one to consider sets that are somewhat dense even though they do not appear to be so when considering only initial segments. This leads us to the concept of (upper) Banach density. In order to defined Banach density, we first need to establish a basic lemma from real analysis, whose nonstandard proof is quite elegant.

**Lemma 10.3 (Fekete)** *Suppose that* $(a_n)$ *is a* subadditive *sequence of positive real numbers, that is,* $a_{m+n} \leq a_m + a_n$ *for all* $m, n$. *Then the sequence* $\left(\frac{1}{n}a_n\right)$ *converges to* $\inf\left\{\frac{1}{n}a_n : n \in \mathbb{N}\right\}$.

*Proof* After normalizing, we may suppose that $a_1 = 1$. This implies that $\frac{1}{n}a_n \leq 1$ for every $n \in \mathbb{N}$. Set $\ell := \inf\{\frac{1}{n}a_n \,:\, n \in \mathbb{N}\}$. By transfer, there exists $v_0 \in {}^*\mathbb{N}$ infinite such that $\frac{1}{v_0}a_{v_0} \approx \ell$. Furthermore $\mathrm{st}\left(\frac{1}{v}a_v\right) \geq \ell$ for every $v \in {}^*\mathbb{N}$. Fix an infinite $\mu \in {}^*\mathbb{N}$ and observe that for $v \geq \mu v_0$ one can write $v = r v_0 + s$ where $r \geq \mu$ and $s < v_0$. Therefore

$$\frac{1}{v}a_v \leq \frac{r a_{v_0} + a_s}{r v_0 + s} \leq \frac{a_{v_0}}{v_0} + \frac{a_s}{\mu s} \leq \frac{a_{v_0}}{v_0} + \frac{1}{\mu} \approx \frac{a_{v_0}}{v_0} \approx \ell.$$

It follows that $\frac{1}{v}a_v \approx \ell$ for every $v \geq \mu v_0$, whence by transfer we have that, for every $\varepsilon > 0$, there exists $n_0 \in \mathbb{N}$ such that $\left|\frac{1}{n}a_n - \ell\right| < \varepsilon$ for every $n \geq n_0$. Therefore the sequence $\left(\frac{1}{n}a_n\right)$ converges to $\ell$.

For each $n$, set

$$\Delta_n(A) := \max\{\delta(A, I) \ : \ I \subseteq \mathbb{N} \text{ is an interval of length } n\}.$$

It is straightforward to verify that $(\Delta_n(A))$ is subadditive, whence, by Fekete's Lemma, we have that the sequence $(\Delta_n(A))$ converges to $\inf_n \Delta_n(A)$.

**Definition 10.4** We define the *Banach density of A* to be

$$BD(A) = \lim_{n \to \infty} \Delta_n(A) = \inf_n \Delta_n(A).$$

*Remark 10.5* Unlike upper and lower densities, the notion of Banach density actually makes sense in any amenable (semi)group, although we will not take up this direction in this book.

If $(I_n)$ is a sequence of intervals in $\mathbb{N}$ such that $\lim_{n\to\infty} |I_n| = \infty$ and $BD(A) = \lim_{n\to\infty} \delta(A, I_n)$, then we say that $(I_n)$ *witnesses the Banach density of A*.

Here is the nonstandard characterization of Banach density:

**Exercise 10.6** For any $N \in {}^*\mathbb{N}\backslash\mathbb{N}$, we have

$$BD(A) = \max\{st(\delta({}^*A, I)) \ : \ I \subseteq {}^*\mathbb{N} \text{ is an interval of length } N\}.$$

As above, if $I$ is an infinite hyperfinite interval such that $BD(A) = st(\delta(A, I))$, we also say that $I$ *witnesses the Banach density of A*.

**Exercise 10.7** Give an example of a set $A \subseteq \mathbb{N}$ such that $\overline{d}(A) = 0$ but $BD(A) = 1$.

**Exercise 10.8** Prove that Banach density is translation-invariant: $BD(A + n) = BD(A)$, where $A + n = \{a + n : a \in A\}$.

Banach density is also subadditive:

**Proposition 10.9** *For any $A, B \subseteq \mathbb{N}$, we have $BD(A \cup B) \leq BD(A) + BD(B)$.*

*Proof* Let $I$ be an infinite hyperfinite interval witnessing the Banach density of $A \cup B$. Then

$$BD(A \cup B) = st(\delta(A \cup B, I)) \leq st(\delta(A, I)) + st(\delta(B, I)) \leq BD(A) + BD(B).$$

The following "fattening" result is often useful.

**Proposition 10.10** *If $BD(A) > 0$, then $\lim_{k \to \infty} BD(A + [-k, k]) = 1$.*

*Proof* Set $r := BD(A)$. For each $k$, set $a_k := \max_{x \in \mathbb{N}} |A \cap [x + 1, x + k]|$, so $r = \lim_{k\to\infty} a_k/k$. By the Squeeze Theorem, it suffices to show that $BD(A + [-k, k]) \geq \frac{r \cdot k}{a_k}$ for all $k$. Towards this end, fix $k \in \mathbb{N}$ and $N \in {}^*\mathbb{N}\backslash\mathbb{N}$ and take $x \in {}^*\mathbb{N}$ such that $s := |{}^*A \cap [x + 1, x + N \cdot k]|/N \cdot k \approx r$. For $i = 0, 1, \ldots, N - 1$, set $J_i := [x + ik + 1, x + (i + 1)k]$. Set $\Lambda := \{i \mid {}^*A \cap J_i \neq \emptyset\}$; observe that $\Lambda$ is internal. We then have

$$s = \frac{|{}^*A \cap [x+1, x+N \cdot k]|}{N \cdot k} = \frac{\sum_{i \in \Lambda} |{}^*A \cap J_i|}{N \cdot k} \leq \frac{|\Lambda| \cdot a_k}{N \cdot k},$$

whence we can conclude that $|\Lambda| \geq s \cdot N \cdot k / a_k$. Now note that if $i \in \Lambda$, then $J_i \subseteq {}^*A + [-k, k]$, so

$$\frac{|({}^*A + [-k, k]) \cap [x+1, x+N \cdot k]|}{N \cdot k} \geq \frac{|\Lambda| \cdot k}{N \cdot k} \geq s \cdot k / a_k.$$

It follows that $BD(A + [-k, k]) \geq r \cdot k / a_k$.

## 10.2 Structural Properties

We now move on to consider structural notions of largeness. In this section, $A$ continues to denote a subset of $\mathbb{N}$.

**Definition 10.11** $A$ is *thick* if and only if $A$ contains arbitrarily long intervals.

**Proposition 10.12** *$A$ is thick if and only if there is an infinite hyperfinite interval $I$ contained in ${}^*A$.*

*Proof* The backwards direction follows directly from transfer. The forwards direction follows from the overflow principle applied to the internal set $\{\alpha \in {}^*\mathbb{N} : {}^*A$ contains an interval of length $\alpha\}$.

**Corollary 10.13** *$A$ is thick if and only if $BD(A) = 1$.*

*Proof* The forwards direction is obvious. For the backwards direction, let $N \in {}^*\mathbb{N}$ be divisible by all elements of $\mathbb{N}$ and let $I$ be a hyperfinite interval of length $N$ witnessing the Banach density of $A$. If $A$ is not thick, then there is $m$ such that $m \mid N$ and $A$ does not contain any intervals of length $m$. Divide $I$ into $N/m$ many intervals of length $m$. By transfer, each such interval contains an element of ${}^*\mathbb{N} \setminus {}^*A$. Thus

$$BD(A) = st(\delta(A, I)) \leq st\left(\frac{N - N/m}{N}\right) = 1 - 1/m.$$

**Definition 10.14** $A$ is *syndetic* if $\mathbb{N} \setminus A$ is not thick.

Equivalently, $A$ is syndetic if there is $m$ such that all gaps of $A$ are of size at most $m$.

**Proposition 10.15** *$A$ is syndetic if and only if all gaps of ${}^*A$ are finite.*

*Proof* The forward direction is immediate by transfer. For the backwards direction, consider the set

$$X := \{\alpha \in {}^*\mathbb{N} : \text{all gaps of } {}^*A \text{ are of size at most } \alpha\}.$$

By assumption, $X$ contains all elements of ${}^*\mathbb{N}\backslash\mathbb{N}$, so by underflow, there is $m \in X \cap \mathbb{N}$. In particular, all gaps of $A$ are of size at most $m$.

**Definition 10.16** $A$ is *piecewise syndetic* if there is a finite set $F \subseteq \mathbb{N}$ such that $A + F$ is thick.

**Proposition 10.17** *If $A$ is piecewise syndetic, then* $\mathrm{BD}(A) > 0$. *More precisely, if $F$ is a finite set such that $A + F$ is thick, then* $\mathrm{BD}(A) \geq 1/|F|$.

*Proof* Take finite $F \subseteq \mathbb{N}$ such that $A + F$ is thick. Since Banach density is translation invariant, by Proposition 10.9, we have

$$1 = \mathrm{BD}(\mathbb{N}) = \mathrm{BD}(\bigcup_{x \in F}(A + x)) \leq |F| \cdot \mathrm{BD}(A).$$

The notion of being piecewise syndetic is very robust in that it has many interesting reformulations:

**Proposition 10.18** *For $A \subseteq \mathbb{N}$, the following are equivalent:*

1. *$A$ is piecewise syndetic;*
2. *there is $m \in \mathbb{N}$ such that $A + [0, m]$ is thick;*
3. *there is $k \in \mathbb{N}$ such that for every $N > \mathbb{N}$, there is a hyperfinite interval $I$ of length $N$ such that ${}^*A$ has gaps of size at most $k$ on $I$;*
4. *for every $N > \mathbb{N}$, there is a hyperfinite interval $I$ of length $N$ such that all gaps of ${}^*A$ on $I$ are finite;*
5. *there is $k \in \mathbb{N}$ and there is an infinite hyperfinite interval $I$ such that ${}^*A$ has gaps of size at most $k$ on $I$;*
6. *there is an infinite hyperfinite interval $I$ such that all gaps of ${}^*A$ on $I$ are finite;*
7. *there is $k \in \mathbb{N}$ such that, for every $n \in \mathbb{N}$, there is an interval $I \subseteq \mathbb{N}$ of length $n$ such that the gaps of $A$ on $I$ are of size at most $k$;*
8. *there is a thick set $B$ and a syndetic set $C$ such that $A = B \cap C$.*

*Proof* Clearly (1) and (2) are equivalent and (3) implies (4). Now assume that (3) fails. In particular, if $X$ is the set of $k \in {}^*\mathbb{N}$ for which there is a hyperfinite interval $I$ of length greater than $k$ on which ${}^*A$ has gaps of size greater than $k$, then $X$ contains all standard natural numbers. By , there is an infinite element of $X$, whence (4) fails. Thus, (3) and (4) are equivalent. (5) clearly implies (6) and (6) implies (5) follows from a familiar underflow argument. (5) and (7) are also equivalent by transfer-overflow.

We now show (2) implies (3). Fix $N > \mathbb{N}$. By (2) and transfer, there is an interval $[x, x + N) \subseteq {}^*A + [0, m]$. Thus, on $[x, x + N)$, ${}^*A$ has gaps of size at most $m$.

Clearly (3) $\Rightarrow$ (5). Now suppose that (5) holds. Choose $k \in \mathbb{N}$ and $M, N \in {}^*\mathbb{N}$ such that $M < N$ and $N - M > \mathbb{N}$ such that ${}^*A$ has gaps of size at most $k$ on $[M, N]$. Then $[M + k, N] \subseteq {}^*A + [0, k]$. It follows by transfer that $A + [0, k]$ is thick, whence (2) holds.

Thus far, we have proven that (1)–(7) are equivalent. Now assume that (7) holds and take $k \in \mathbb{N}$ and intervals $I_n \subseteq \mathbb{N}$ of length $n$ such that $A$ has gaps of size at most $k$ on each $I_n$. Without loss of generality, the $I_n$'s are of distance at least $k + 1$ from each other. Let $B := A \cup \bigcup_n I_n$ and let $C := A \cup (\mathbb{N} \backslash B)$. Clearly $B$ is thick. To see that $C$ is syndetic, suppose that $J$ is an interval of size $k + 1$ disjoint from $C$. Then $J$ is disjoint from $A$ and $J \subseteq B$, whence $J \subseteq \bigcup_n I_n$. Since the $I_n$'s are of distance at least $k + 1$ from each other, $J \subseteq I_n$ for some $n$. Thus, $J$ represents a gap of $A$ on $I_n$ of size $k + 1$, yielding a contradiction. It is clear that $A = B \cap C$.

Finally, we prove that (8) implies (7). Indeed, suppose that $A = B \cap C$ with $B$ thick and $C$ syndetic. Suppose that $k \in \mathbb{N}$ is such that all gaps of $C$ are of size at most $k$. Fix $n \in \mathbb{N}$ and let $I$ be an interval of length $n$ contained in $B$. If $J$ is an interval contained in $I$ of size $k + 1$, then $J \cap C \neq \emptyset$, whence $J \cap A \neq \emptyset$ and (7) holds.

Item (7) in the previous proposition explains the name piecewise syndetic. The following is not obvious from the definition:

**Corollary 10.19** *The notion of being piecewise is partition regular, meaning that if $A$ is piecewise syndetic and $A = A_1 \sqcup A_2$, then $A_i$ is piecewise syndetic for some $i = 1, 2$.*

*Proof* Suppose that $I$ is an infinite hyperfinite interval such that all gaps of ${}^*A$ on $I$ are finite. Suppose that $I$ does not witness that $A_1$ is piecewise syndetic. Then there is an infinite hyperfinite interval $J \subseteq I$ such that $J \cap {}^*A_1 = \emptyset$. It then follows that any gap of ${}^*A_2$ on $J$ must be finite, whence $J$ witnesses that $A_2$ is piecewise syndetic.

*Remark 10.20* We note that neither thickness nor syndeticity are partition regular notions. Indeed, if $A$ is the set of even numbers and $B$ is the set of odd numbers, then neither $A$ nor $B$ is thick but their union certainly is. For syndeticity, let $(x_n)$ be the sequence defined by $x_1 = 1$ and $x_{n+1} := x_n + n$. Set $C := \bigcup_{n \text{ even}} [x_n, x_n + n)$ and $D := \bigcup_{n \text{ odd}} [x_n, x_n + n)$. Then neither $C$ nor $D$ are syndetic but their union is $\mathbb{N}$, a syndetic set.

The following is a nice consequence of the partition regularity of the notion of piecewise syndetic.

**Corollary 10.21** *van der Waerden's theorem is equivalent to the statement that piecewise syndetic sets contain arbitrarily long arithmetic progressions.*

*Proof* First suppose that van der Waerden's theorem holds and let $A$ be a piecewise syndetic set. Fix $k \in \mathbb{N}$; we wish to show that $A$ contains an arithmetic progression of length $k$. Take $m$ such that $A + [0, m]$ is thick. Let $l$ be sufficiently large such that when intervals of length $l$ are partitioned into $m + 1$ pieces, then there is a monochromatic arithmetic progression of length $k$. Let $I \subseteq A+[0, m]$ be an interval of length $l$. Without loss of generality, we may suppose that the left endpoint of $I$ is greater than $m$. Let $c$ be the coloring of $I$ given by $c(x) :=$ the least $i \in [0, m]$ such that $x \in A+i$. Then there is $i \in [0, m]$ and $x, d$ such that $x, x+d, \ldots, x+(k-1)d \in A + i$. It follows that $(x - i), (x - i) + d, \ldots, (x - i) + (k - 1)d \in A$.

Conversely, suppose that piecewise syndetic sets contain arbitrarily long arithmetic progressions. Fix a finite coloring $c$ of the natural numbers. Since being piecewise syndetic is partition regular, some color is piecewise syndetic, whence contains arbitrarily long arithmetic progressions by assumption.

## 10.3 Working in $\mathbb{Z}$

We now describe what the above densities and structural properties mean in the group $\mathbb{Z}$ as opposed to the semigroup $\mathbb{N}$. Thus, in this section, $A$ now denotes a subset of $\mathbb{Z}$.

It is rather straightforward to define the appropriate notions of density. Indeed, given any sequence $(I_n)$ of intervals in $\mathbb{Z}$ with $\lim_{n \to \infty} |I_n| = \infty$, we define

$$\overline{d}_{(I_n)} := \limsup_{n \to \infty} \delta(A, I_n)$$

and

$$\underline{d}_{(I_n)} := \liminf_{n \to \infty} \delta(A, I_n).$$

When $I_n = [-n, n]$ for each $n$, we simply write $\overline{d}(A)$ (resp. $\underline{d}(A)$) and speak of the *upper* (resp. *lower*) density of $A$. Finally, we define the *upper Banach density of $A$* to be

$$BD(A) = \lim_{n \to \infty} \max_{x \in \mathbb{N}} \delta(A, [x - n, x + n]).$$

Of course, one must verify that this limit exists, but this is proven in the exact same way as in the case of subsets of $\mathbb{N}$.

**Exercise 10.22** Prove that

$$BD(A) := \max\{\overline{d}_{(I_n)}(A) \, : \, (I_n) \text{ a sequence of intervals with } \lim_{n \to \infty} |I_n| = \infty\}.$$

The notions of thickness and syndeticity for subsets of $\mathbb{Z}$ remains unchanged: $A$ is thick if $A$ contains arbitrarily long intervals and $A$ is syndetic if $\mathbb{Z} \setminus A$ is not thick.

Similarly, $A$ is piecewise syndetic if there is a finite set $F \subseteq \mathbb{Z}$ such that $A + F$ is thick. The following lemma is almost immediate:

**Lemma 10.23** *$A$ is piecewise syndetic if and only if there is a finite set $F \subseteq \mathbb{Z}$ such that, for every finite $L \subseteq \mathbb{Z}$, we have $\bigcap_{x \in L}(A + F + x) \neq \emptyset$.*

**Exercise 10.24** Formulate and verify all of the nonstandard equivalents of the above density and structural notions developed in the previous two sections for subsets of $\mathbb{Z}$.

The following well-known fact about difference sets has a nice nonstandard proof.

**Proposition 10.25** *Suppose that $A \subseteq \mathbb{Z}$ is such that $BD(A) > 0$. Then $A - A$ is syndetic. In fact, if $BD(A) = r$, then there is a finite set $F \subseteq \mathbb{Z}$ with $|F| \leq \frac{1}{r}$ such that $(A - A) + F = \mathbb{Z}$.*

First, we need a lemma.

**Lemma 10.26** *Let $N \in {}^*\mathbb{N} \backslash \mathbb{N}$. Suppose that $E \subseteq [1, N]$ is an internal set such that $\delta(E, N) \approx r$. Then there is a finite $F \subseteq \mathbb{Z}$ with $|F| \leq 1/r$ such that $\mathbb{Z} \subseteq (E - E) + F$.*

*Proof* Fix $x_1 \in \mathbb{N}$. If $\mathbb{Z} \subseteq (E - E) + x_1$, then take $F = \{x_1\}$. Otherwise, take $x_2 \notin (E - E) + \{x_1\}$. If $\mathbb{Z} \subseteq (E - E) + \{x_1, x_2\}$, then take $F = \{x_1, x_2\}$. Otherwise, take $x_3 \notin (E - E) + \{x_1, x_2\}$.

Suppose that $x_1, \ldots, x_k$ have been constructed in this fashion. Note that the sets $E + x_i$, for $i = 1, \ldots, k$, are pairwise disjoint. Since each $x_i \in \mathbb{Z}$ and $N$ is infinite, we have that $\delta((E + x_i), N) \approx r$. It follows that

$$\delta\left(\bigcup_{i=1}^{k}(E + x_i), N\right) = \frac{\sum_{i=1}^{k}|(E + x_i) \cap [1, N]|}{N} \approx kr.$$

It follows that the process must stop after $k$-many steps, with $k \leq \frac{1}{r}$.

*Proof (of Proposition 10.25)* Set $r := BD(A)$. Fix and infinite $N$ and take $x \in {}^*\mathbb{N}$ such that $\delta({}^*A, [x+1, x+N]) \approx r$. Set $E := ({}^*A - x) \cap [1, N]$. Then $\delta(E, N) \approx r$, whence there is finite $F \subseteq \mathbb{Z}$ with $|F| \leq 1/r$ such that $\mathbb{Z} \subseteq (E - E) + F$. It follows that $\mathbb{Z} \subseteq ({}^*A - {}^*A) + F$, whence it follows by transfer that $\mathbb{Z} = (A - A) + F$.

The analog of Proposition 10.10 for $\mathbb{Z}$ is also true:

**Proposition 10.27** *If $BD(A) > 0$, then $\lim_{k \to \infty} BD(A + [-k, k]) = 1$.*

However, for our purposes in Sect. 12.5, we will need a more precise result. Note that, a priori, for every $\epsilon > 0$, there is $k_\epsilon$ and infinite hyperfinite interval $I_\epsilon$ such that $\delta({}^*A + [-k_\epsilon, k_\epsilon], I_\epsilon) > 1 - \epsilon$. The next proposition tells us that we can take a single interval $I$ to work for each $\epsilon$. The proof is heavily inspired by the proof of [8, Lemma 3.2].

**Proposition 10.28** *Suppose that* BD$(A) > 0$. *Then there is an infinite hyperfinite interval* $I \subseteq \mathbb{Z}$ *such that, for every* $\epsilon > 0$, *there is $k$ for which* $\delta(^*A + [-k, k], I) > 1 - \epsilon$.

*Proof* Pick $(I_n)$ a sequence of intervals in $\mathbb{Z}$ witnessing the Banach density of $A$ and such that, for every $k$, we have that $\lim_{n\to\infty} \delta(A + [-k, k], I_n)$ exists. (This is possible by a simple diagonalization argument.) Fix an infinite $N$ and, for each $\alpha \in {}^*\mathbb{N}$, set $G_\alpha := (^*A + [-\alpha, \alpha]) \cap I_N$. Set $r := \sup_{k\in\mathbb{N}} \mu_{I_N}(G_k)$.

*Claim* There is $K > N$ such that:

(i) For every $l \in \mathbb{Z}$, $\frac{|(l+G_K)\Delta G_K|}{|G_K|} \approx 0$.

(ii) $\frac{|G_K|}{|I_N|} \approx r$.

*Proof of Claim* For each $l \in \mathbb{Z}$, set $X_l$ to be the set of $\alpha \in {}^*\mathbb{N}$ such that:

(a) $\alpha \geq l$;
(b) For all $x \in \mathbb{Z}$ with $|x| \leq l$, we have $\frac{|(x+G_\alpha)\Delta G_\alpha|}{|G_\alpha|} < \frac{1}{l}$;
(c) $\left| \frac{|G_\alpha|}{|I_N|} - r \right| < \frac{1}{l}$.

Since each $X_l$ is internal and unbounded in $\mathbb{N}$, by saturation there is $K \in \bigcap_l X_l$. This $K$ is as desired.

Fix $K$ as in the Claim and set $G := G_K$ and $\mu := \mu_G$. For $k \in \mathbb{N}$, we then have that

$$\delta(^*A + [-k, k], G) = \frac{|(^*A + [-k, k]) \cap I_N|}{|G|} \approx \delta(^*A + [-k, k], I_N) \cdot \frac{1}{r},$$

whence we see that $\delta(^*A + [-k, k], G) \to 1$ as $k \to \infty$.[1]

Now take $J$ to be an infinite hyperfinite interval such that $\frac{|(l+G_K)\Delta G_K|}{|G_K|} \approx 0$ for all $l \in J$; this is possible as a consequence of the overflow principle. We claim that there is $t \in G$ such that $I := t + J$ is as desired.

For each $k$, take $n_k$ such that $\delta(^*A + [-n_k, n_k], G) > 1 - \frac{1}{k}$; without loss of generality, we may assume that $(n_k)$ is an increasing sequence. Set $B_k := {}^*A + [-n_k, n_k]$ and set $g_k : G \to [0, 1]$ to be the $\mathscr{L}_G$-measurable function given by $g_k(t) := \text{st}(\delta(B_k, t + J))$. For each $t \in G$, we have that $(g_k(t))$ is a bounded nondecreasing sequence, whence converges to a limit $g(t)$. By the Dominated Convergence Theorem, we have that $\int_G g(t)d\mu = \lim_{k\to\infty} \int_G g_k(t)d\mu$.

---

[1] At this point, we may note that $G$ satisfies the conclusion of the proposition except that it is not an interval but instead a *Følner approximation* for $\mathbb{Z}$. While this would suffice for our purposes in Sect. 12.5, we wanted to avoid having to introduce the theory of Følner approximations and instead opted to work a bit harder to obtain the above cleaner statement.

Now note that

$$\int_G g_k(t)d\mu \approx \frac{1}{|G|} \sum_{t \in G} \delta(B_k, t + J) = \frac{1}{|I|} \sum_{x \in J} \delta(B_k, x + G) \approx \delta(B_k, G) > 1 - \frac{1}{k}.$$

It follows that $\int_G g(t)d\mu = 1$, whence $g(t) = 1$ for some $t \in G$. It is then clear that $I := t + J$ is as desired.

We call $I$ as in the conclusion of Proposition 10.28 *good for A*. One can also prove the previous proposition using a Lebesgue Density Theorem for cut spaces; see [41].

## 10.4  Furstenberg's Correspondence Principle

We end this chapter by explaining the nonstandard take on Furstenberg's correspondence principle.

**Theorem 10.29 (Furstenberg's Correspondence Principle)** *Suppose that $A \subseteq \mathbb{Z}$ is such that $BD(A) > 0$. Then there is a measure-preserving dynamical system $(X, \mathcal{B}, \nu, T)$ and a measurable set $A_0 \in \mathcal{B}$ such that $\nu(A_0) = BD(A)$ and such that, for any finite set $F \subseteq \mathbb{Z}$, we have:*

$$BD \left( \bigcap_{i \in F} (A - i) \right) \geq \nu \left( \bigcap_{i \in F} T^{-i}(A_0) \right).$$

*Proof* Fix $I \subseteq {}^*\mathbb{Z}$ witnessing the Banach density of $A$. It is easy to verify that the hypercycle system $(I, \Omega, \mu, S)$ introduced in Sect. 5.6 of Chap. 5 and the set $A_0 := {}^*A \cap I$ are as desired.

Let us mention the ergodic-theoretic fact that Furstenberg proved:

**Theorem 10.30 (Furstenberg Multiple Recurrence Theorem)** *Suppose that $(X, \mathcal{B}, \nu, T)$ is a measure-preserving dynamical system, $A \in \mathcal{B}$ is such that $\nu(A) > 0$, and $k \in \mathbb{N}$ is given. Then there exists $n \in \mathbb{N}$ such that $\nu(A \cap T^{-n}(A) \cap T^{-2n}(A) \cap \cdots \cap T^{-(k-1)n}(A)) > 0$.*

Notice that the above theorem, coupled with the Furstenberg Correspondence Principle, yields Furstenberg's proof of Szemerédi's Theorem .

**Theorem 10.31 (Szemeredi's Theorem)** *If $A \subseteq \mathbb{Z}$ is such that $BD(A) > 0$, then $A$ contains arbitrarily long arithmetic progressions.*

Szemeredi's Theorem is the density version of van der Waerden's theorem and was originally proven by Szemeredi in [119].

We end this chapter giving a simpler application of the correspondence principle used by Bergelson in [12] to give a quantitative version of Schur's Theorem.

Suppose that $c : \mathbb{N} \to \{1, \ldots, m\}$ is an $m$-coloring of $\mathbb{N}$. Then Schur's theorem states that there is $i \in \{1, \ldots, m\}$ and $a, b \in \mathbb{N}$ such that $c(a) = c(b) = c(a + b) = i$. (Note that Schur's theorem is an immediate corollary of Rado's Theorem.) It is natural to ask whether or not a quantitative Schur's theorem could hold in the sense that there should be some color $C_i$ such that there are many $a, b \in \mathbb{N}$ with $c(a) = c(b) = c(a+b) = i$. In [12], Bergelson proved the following precise version of that result:

**Theorem 10.32** *Suppose that* $c : \mathbb{N} \to \{1, \ldots, m\}$ *is an m-coloring of* $\mathbb{N}$ *and* $C_i := \{n \in \mathbb{N} : c(n) = i\}$. *For* $i \in \{1, \ldots, n\}$ *and* $\epsilon > 0$, *set*

$$R_{i,\epsilon} := \{n \in C_i : \overline{d}(C_i \cap (C_i - n)) \geq \overline{d}(C_i)^2 - \epsilon\}.$$

*Then there is* $i \in \{1, \ldots, n\}$ *such that, for every* $\epsilon > 0$, *we have* $\overline{d}(R_{i,\epsilon}) > 0$.

We more or less follow Bergelson's original proof except we use the nonstandard version of the Furstenberg correspondence principle.

**Definition 10.33** We call $R \subseteq \mathbb{N}$ a *set of nice recurrence* if: given any dynamical system $(X, \mathscr{B}, \mu, T)$, any $\mu(B) > 0$, and any $\epsilon > 0$, there is $n \in R$ such that $\mu(A \cap T^{-n}A) \geq \mu(A)^2 - \epsilon$.

**Proposition 10.34** *Let* $S \subseteq \mathbb{N}$ *be an infinite set. Then* $S - S$ *is a set of nice recurrence.*

*Proof* Let $(s_i)$ be an enumeration of $S$ in increasing order. It is straightforward to check that there must exist $i < j$ such that $\mu(T^{-s_i}A \cap T^{-s_j}A) \geq \mu(A)^2 - \epsilon$. It follows that $\mu(A \cap T^{-(s_j - s_i)}A) \geq \mu(A)^2 - \epsilon$, as desired.

**Exercise 10.35** If $E \subseteq \mathbb{N}$ is thick, then there is an infinite set $S \subseteq \mathbb{N}$ such that $S - S \subseteq E$.

**Corollary 10.36** *Suppose that* $E \subseteq \mathbb{N}$ *is thick and* $E = C_1 \cup \cdots \cup C_k$ *is a partition of* $E$. *Then some* $C_i$ *is a set of nice recurrence.*

*Proof* By Exercise 10.35, we may take $S \subseteq \mathbb{N}$ such that $S - S \subseteq E$. Define a coloring $c : S \to \{1, \ldots, k\}$ by declaring, for $s, s' \in S$ with $s < s'$, that $c(\{s, s'\}) := i$ if $c(s - s') = i$. By Ramsey's theorem, there is an infinite $S' \subseteq S$ and $i \in \{1, \ldots, k\}$ such that $c([S']^2) = \{i\}$. It follows that $S' - S' \subseteq C_i$. By Proposition 10.34, $S' - S'$, and hence $C_i$, is a nice set of recurrence.

We are now ready to give the proof of Theorem 10.32. First, without loss of generality, we may assume that there is $k \in \{1, \ldots, m\}$ such that $\overline{d}(C_i) > 0$ for $i = 1, \ldots, k$ and $C_1 \cup \cdots \cup C_k$ is thick. For ease of notation, for $p \in \mathbb{N}$, let $R_{i,p} := R_{i,1/p}$. It suffices to show that, for each $p \in \mathbb{N}$, there is $i_p \in \{1, \ldots, k\}$ such that $\overline{d}(R_{i_p,p}) > 0$. Indeed, if this is the case, then by the Pigeonhole Principle, there is some $i \in \{1, \ldots, m\}$ such that $i_p = i$ for infinitely many $p$; this $i$ is as desired.

Towards this end, fix $p \in \mathbb{N}$ and, again for ease of notation, set $R_i := R_{i,p}$. Suppose, towards a contradiction, that $\overline{d}(R_i) = 0$ for each $i = 1, \ldots, k$. Set $D_i := C_i \backslash R_i$. Then $\overline{d}(D_i) = \overline{d}(C_i)$ and $D_1 \cup \cdots D_k$ is thick. By Corollary 10.36, there is $i \in \{1, \ldots, k\}$ such that $D_i$ is a nice set of recurrence. Take $N > \mathbb{N}$ such that $\overline{d}(D_i) = \mu_N(^*D_i)$. By applying the fact that $D_i$ is a nice set of recurrence to the hypercycle system based on $[1, N]$ and the measurable set $A := {}^*D_i \cap [1, N]$, we get that there is $n \in D_i$ such that

$$\overline{d}(C_i \cap (C_i - n)) \geq \overline{d}(D_i \cap (D_i - n)) \geq \mu_N(A \cap T^{-n}A) \geq \mu(A)^2 - \epsilon = \overline{d}(C_i)^2 - \epsilon,$$

contradicting the fact that $n \notin R_i$.

## Notes and References

The first appearance of nonstandard methods in connection with densities and structural properties seems to be Leth's dissertation and subsequent article [88]. Proposition 10.10 was first proven by Hindman in [70]. Partition regularity of piecewise syndeticity was first proven by Brown in [21]. Proposition 10.25 was first proven by Følner in [49]; the nonstandard proof is due to Di Nasso [36]. Furstenberg's Correspondence Principle was first established in [52] where he gave his ergodic-theoretic proof of Szemerédi's theorem. The nonstandard approach to the Furstenberg Correspondence Principle seems to have a somewhat nebulous history. Indeed, while it was surely known to many experts that one could use hypercycle systems to prove the Furstenberg Correspondence Principle, the first appearance of this idea in the literature seems to be generalizations of the Furstenberg Correspondence due to Townser appearing in the paper [125].

# Chapter 11
# Working in the Remote Realm

In this chapter, we present a technique, due to Jin, of converting some theorems about sets of positive upper asymptotic density or Shnirelmann density (to be defined below) to analogous theorems about sets of positive Banach density. The novel idea is to use the ergodic theorem for hypercycles applied to characteristic functions of nonstandard extensions. Deviating from Jin's original formulation, we present his technique using the more recent notion of finite embeddability of subsets of natural numbers due to Di Nasso.

## 11.1 Finite Embeddability Between Sets of Natural Numbers

A useful combinatorial notion is the following:

**Definition 11.1** Let $X, Y$ be sets of integers. We say that $X$ is *finitely embeddable* in $Y$, and write $X \lhd Y$, if every finite configuration $F \subseteq X$ has a shifted copy $t + F \subseteq Y$.

Finite embeddability preserves most of the fundamental combinatorial notions that are commonly considered in combinatorics of integer numbers.

**Proposition 11.2**

1. *A set is $\lhd$-maximal if and only if it is thick.*
2. *If $X$ contains an arithmetic progression of length $k$ and distance $d$ and $X \lhd Y$, then also $Y$ also contains an arithmetic progression of length $k$ and distance $d$.*
3. *If $X$ is piecewise syndetic and $X \lhd Y$, then also $Y$ is piecewise syndetic.*
4. *If $X \lhd Y$, then $\mathrm{BD}(X) \leq \mathrm{BD}(Y)$.*

© Springer Nature Switzerland AG 2019
M. Di Nasso et al., *Nonstandard Methods in Ramsey Theory and Combinatorial Number Theory*, Lecture Notes in Mathematics 2239,
https://doi.org/10.1007/978-3-030-17956-4_11

*Proof* (1) Clearly $X$ is maximal if and only if $\mathbb{N} \lhd X$ if and only if every finite interval $[1, n]$ has a shifted copy $[x + 1, x + n] \subseteq X$. (2) is trivial. We leave the proofs of (3) and (4) to the reader.

We stress the fact that while piecewise syndeticity is preserved under finite embeddability, the property of being syndetic is not. Similarly, the upper Banach density is preserved or increased under finite embeddability, but the upper asymptotic density is not. A list of basic properties is itemized below.

**Proposition 11.3**

1. *If $X \lhd Y$ and $Y \lhd Z$, then $X \lhd Z$.*
2. *If $X \lhd Y$ and $X' \lhd Y'$, then $X - X' \lhd Y - Y'$.*
3. *If $X \lhd Y$, then $\bigcap_{t \in G}(X - t) \lhd \bigcap_{t \in G}(Y - t)$ for every finite $G$.*

*Proof*

(1) is straightforward from the definition of $\lhd$.

(2). Given a finite $F \subseteq X - X'$, let $G \subseteq X$ and $G' \subseteq X'$ be finite sets such that $F \subseteq G - G'$. By the hypotheses, there exist $t, t'$ such that $t + G \subseteq Y$ and $t' + G' \subseteq Y'$. Then, $(t - t') + F \subseteq (t + G) - (t' + G') \subseteq Y - Y'$.

(3). Let a finite set $F \subseteq \bigcap_{t \in G}(X - t)$ be given. Notice that $F + G \subseteq X$, so we can pick an element $w$ such that $w + (F + G) \subseteq Y$. Then, $w + F \subseteq \bigcap_{t \in G} Y - t$.

**Definition 11.4** Let $A \subseteq \mathbb{N}$ and $a \in {}^*\mathbb{N}$. The *remote realm* of $A$ at $\alpha$ is the following subset of ${}^*\mathbb{N}$:

$$A_\alpha := ({}^*A - \alpha) \cap \mathbb{N} = \{n \in \mathbb{N} \mid a + n \in {}^*A\}.$$

In a nonstandard setting, the finite embeddability $X \lhd Y$ means that $X$ is contained in some remote realm of $Y$. This notion can be also characterized in terms of *ultrafilter-shifts*, as defined by Beiglböck [7].

**Proposition 11.5** *Let $X, Y \subseteq \mathbb{N}$. Then the following are equivalent:*

1. *$X \lhd Y$.*
2. *There exists $a \in {}^*\mathbb{N}$ such that $X \subseteq Y_a$ (that is, $a + X \subseteq {}^*Y$).*
3. *There exists an ultrafilter $\mathscr{U}$ on $\mathbb{N}$ such that $X \subseteq Y - \mathscr{U}$, where the ultrafilter shift $Y - \mathscr{U} := \{x : Y - x \in \mathscr{U}\}$.*

*Proof* (1) $\Rightarrow$ (2). Enumerate $X = \{x_n \mid n \in \mathbb{N}\}$. By the hypothesis, the finite intersection $\bigcap_{i=1}^n (Y - x_i) \neq \emptyset$. Then, by *overspill*, there exists an infinite $N \in {}^*\mathbb{N}$ such that $\bigcap_{i=1}^N ({}^*Y - x_i)$ is non-empty. If $a \in {}^*\mathbb{N}$ is in that intersection, then clearly $a + x_i \in {}^*Y$ for all $i \in \mathbb{N}$.

(2) $\Rightarrow$ (3). Let $\mathscr{U} = \mathscr{U}_a$ be the ultrafilter generated by $a \in {}^*\mathbb{N}$. For every $x \in X$, by the hypothesis, $a + x \in {}^*Y \Rightarrow a \in {}^*(Y - x)$, and hence $Y - x \in \mathscr{U}$, i.e., $x \in Y - \mathscr{U}$, as desired.

(3) $\Rightarrow$ (1). Given a finite $F \subseteq X$, the set $\bigcap_{x \in F}(Y - x)$ is nonempty, because it is a finite intersection of elements of $\mathscr{U}$. If $t \in \mathbb{Z}$ is any element in that intersection, then $t + F \subseteq Y$.

One can also considers a notion of *dense embeddability* $X \lhd_d Y$ when every finite configuration $F \subseteq X$ has "densely-many" shifted copies included in $Y$, i.e., if the intersection $\bigcap_{x \in F}(Y - x) = \{t \in \mathbb{Z} \mid t + F \subseteq Y\}$ has positive upper Banach density (see [36]).

The notion of finite embeddability has a natural generalization to ultrafilters.

**Definition 11.6** Let $\mathscr{U}$ and $\mathscr{V}$ be ultrafilters on $\mathbb{N}$. We say that $\mathscr{U}$ is *finitely embeddable* in $\mathscr{V}$, and write $\mathscr{U} \lhd \mathscr{V}$, if for every $B \in \mathscr{V}$ there exists $A \in \mathscr{U}$ with $A \lhd B$.

**Proposition 11.7** *Let $\mathscr{U}, \mathscr{V}$ be ultrafilters on $\mathbb{N}$. Then $\mathscr{U} \lhd \mathscr{V}$ if and only if $\mathscr{V}$ belongs to the closure of $\{\mathscr{U} \oplus \mathscr{W} \mid \mathscr{W} \in \beta\mathbb{N}\}$ in $\beta\mathbb{N}$.*

*Proof* First assume that $\mathscr{U}_\alpha \lhd \mathscr{U}_\beta$. We want to show that for every $B \in \mathscr{U}_\beta$ there exists $\gamma \in {}^*\mathbb{N}$ such that $B \in \mathscr{U}_\alpha \oplus \mathscr{U}_\gamma$. By assumption, we can pick $A \in \mathscr{U}_\alpha$ with $A \lhd B$. By the nonstandard characterization of finite embeddability of sets given above, this means that there exists $\gamma \in {}^*\mathbb{N}$ such that $\gamma + A \subseteq {}^*B$. Then, by transfer, ${}^*\gamma + {}^*A \subseteq {}^{**}B$ and so ${}^*\gamma + \alpha \in {}^{**}B$. This gives the desired property that $B \in \mathscr{U}_\alpha \oplus \mathscr{U}_\gamma$.[1]

Conversely, if $\mathscr{U}_\beta$ is in the closure of $\{\mathscr{U}_\alpha \oplus \mathscr{U}_\gamma \mid \gamma \in {}^*\mathbb{N}\}$, then for every $B \in \mathscr{U}_\beta$ there exists $\gamma$ such that $B \in \mathscr{U}_\alpha \oplus \mathscr{U}_\gamma$, that is, $\alpha \in {}^*B_\gamma$. But then we have found a set $B_\gamma \in \mathscr{U}_\alpha$ with $B_\gamma \lhd B$, as desired.

**Theorem 11.8** *Let $\mathscr{V}$ be an ultrafilter on $\mathbb{N}$. Then $\mathscr{V}$ is maximal with respect to $\lhd$ (that is, $\mathscr{U} \lhd \mathscr{V}$ for every $\mathscr{U} \in \beta\mathbb{N}$) if and only if every $B \in \mathscr{V}$ is piecewise syndetic.*[2]

*Proof* Assume first that every $B \in \mathscr{V}$ is piecewise syndetic. By assumption, there a finite union $T := \bigcup_{i=1}^k (B - n_i)$ that is thick. Pick $\gamma \in {}^*\mathbb{N}$ such that $\gamma + \mathbb{N} \subseteq {}^*T$, and set $A_i := \{m \in \mathbb{N} \mid \gamma + m \in {}^*B - n_i\}$. Note that $\mathbb{N} = \bigcup_{i=1}^k A_i$. Fix an arbitrary ultrafilter $\mathscr{U}$ on $\mathbb{N}$. Take $i$ such that $A_i \in \mathscr{U}$. This means that $\gamma + n_i + A_i \subseteq {}^*B$, and we conclude that $A_i \lhd B$. It follows that $\mathscr{U} \lhd \mathscr{V}$, as desired.

Conversely, assume that $\mathscr{V}$ is maximal, and pick any ultrafilter $\mathscr{U}$ that only contains piecewise syndetic sets. By the hypothesis, for every $B \in \mathscr{V}$ there exists $A \in \mathscr{U}$ such that $A \lhd B$. Since $A$ is piecewise syndetic, it follows from Proposition 11.2, that $B$ is also piecewise syndetic.

Several other results about finite embeddability between ultrafilters and its generalizations are found in [96].

---

[1] Recall that by Proposition 4.5 we have that $X \in \mathscr{U}_\alpha \oplus \mathscr{U}_\beta \Leftrightarrow \alpha + {}^*\beta \in {}^{**}B$.

[2] The property that every $B \in \mathscr{V}$ is piecewise syndetic is equivalent to the property that $\mathscr{V}$ belongs to the closure of the smallest ideal $K(\beta\mathbb{N}, \oplus)$ (see [72, Theorem 4.40]).

## 11.2  Banach Density as Shnirelmann Density in the Remote Realm

The title in this chapter refers to looking at copies of $\mathbb{N}$ starting at some infinite element $a \in {}^*\mathbb{N}$ and then connecting some density of the set of points of this copy of $\mathbb{N}$ that lie in the nonstandard extension of a set $A$ and some other density of the original set $A$ itself. In this regard, given $A \subseteq \mathbb{N}$ and $a \in {}^*\mathbb{N}$, we set $\overline{d}({}^*A - a) := \overline{d}(({}^*A - a) \cap \mathbb{N})$ and likewise for other notions of density. We warn the reader that, in general, we do not identify ${}^*A - a$ and $({}^*A - a) \cap \mathbb{N}$ as sets, but since we have not defined the density of a subset of ${}^*\mathbb{N}$, our convention should not cause too much confusion.

The key observation of Renling Jin is that there is a strong converse to item (3) of Proposition 11.2.

**Proposition 11.9** *Suppose that $A \subseteq \mathbb{N}$ is such that $\mathrm{BD}(A) = r$. Let $I$ be an interval of infinite hyperfinite length witnessing the Banach density of $A$. Then for $\mu_I$-almost all $x \in I$, we have $d({}^*A - x) = r$.*

*Proof* Write $I = [H, K]$ and consider the hypercycle system $(I, \mathscr{L}_i, \mu_I, S)$. Let $f$ denote the characteristic function of ${}^*A \cap I$. It follows that, for $x \in I^{\#} := \bigcap_{n \in \mathbb{N}}[H, K - n]$, we have that

$$\frac{1}{n} \sum_{m=0}^{n-1} f(S^m(x)) = \delta({}^*A, [x, x + n - 1]).$$

By the ergodic theorem for hypercycles (Theorem 5.24), there is a $\mathscr{L}_I$-measurable function $\hat{f}$ such that, for $\mu_I$-almost all $x \in I$, we have that

$$\lim_{n \to \infty} \frac{1}{n} \sum_{m=0}^{n-1} f(S^m(x)) = \hat{f}(x).$$

Since $I^{\#}$ is a $\mu_I$-conull set, we will thus be finished if we can show that $\hat{f}$ is $\mu_I$-almost everywhere equal to $r$ on $I^{\#}$.

Towards this end, first note that $\hat{f}(x) \le r$ for $\mu_I$-almost all $x \in I^{\#}$. Indeed, if $\hat{f}(x) > r$ for a positive measure set of $x \in I^{\#}$, then there would be some $x \in I^{\#}$ with $d({}^*A - x) > r$, whence $\mathrm{BD}(A) > r$ by transfer, yielding a contradiction.

Next note that, by the Dominated Convergence Theorem, we have that

$$\int_I \hat{f}(x) d\mu_I = \lim_{n \to \infty} \int_I \frac{1}{n} \sum_{m=0}^{n-1} f(S^m(x)) d\mu_I = r,$$

where the last equality follows from the fact that $S$ is measure-preserving and that $\int_I \hat{f}(x) d\mu_I = \mu_I(^*A) = r$. By a standard measure theory argument, we have that $\hat{f}(x) = r$ for almost all $x \in I^\#$.

**Remark 11.10** In the context of the previous proposition, since $\mu_I(^*A) > 0$, we can conclude that there is $x \in {}^*A$ such that $d(^*A - x) = r$.

Summarizing what we have seen thus far:

**Theorem 11.11** *For $A \subseteq \mathbb{N}$, the following are equivalent:*

1. $BD(A) \geq r$.
2. *There is $B \lhd A$ such that $\underline{d}(A) \geq r$.*
3. *For any infinite hyperfinite interval $I$ witnessing the Banach density of $A$, we have $d(^*A - x) \geq r$ for $\mu_I$-almost all $x \in I$.*

We now introduce a new notion of density.

**Definition 11.12** For $A \subseteq \mathbb{N}$, we define the *Shnirelman density* of $A$ to be

$$\sigma(A) := \inf_{n \geq 1} \delta(A, n).$$

It is clear from the definition that $\underline{d}(A) \geq \sigma(A)$. Note that the Shnirelman density is very sensitive to what happens for "small" $n$. For example, if $1 \notin A$, then $\sigma(A) = 0$. On the other hand, knowing that $\sigma(A) \geq r$ is a fairly strong assumption and thus there are nice structural results for sets of positive Shnirelman density. We will return to this topic in the next section.

A crucial idea of Jin was to add one more equivalence to the above theorem, namely that there is $B \lhd A$ such that $\sigma(B) \geq r$; in this way, one can prove Banach density parallels of theorems about Shnirelman density. To add this equivalence, one first needs a standard lemma.

**Lemma 11.13** *Suppose that $A \subseteq \mathbb{N}$ is such that $\underline{d}(A) = r$. Then for every $\epsilon > 0$, there is $n_0 \in \mathbb{N}$ such that $\sigma(A - n_0) \geq r - \epsilon$.*

*Proof* Suppose that the lemma is false for a given $\epsilon > 0$. In particular, $\sigma(A) < r - \epsilon$, so there is $n_0 \in \mathbb{N}$ such that $\delta(A, n_0) < r - \epsilon$. Since $n_0$ does not witness the truth of the lemma, there is $n_1 \in \mathbb{N}$ such that $\delta((A - n_0), n_1) < r - \epsilon$. Continuing in this way, we find a sequence $(n_i)$ of natural numbers such that, for all $i$, we have $\delta(A, [n_0 + \ldots + n_i + 1, n_0 + \ldots + n_i + n_{i+1}]) < r - \epsilon$ for all $i$. In consequence, the increasing sequence $(\sum_{i \leq k} n_i)$ witnesses that $\underline{d}(A) \leq r - \epsilon$, yielding a contradiction.

**Proposition 11.14** *Suppose that $BD(A) \geq r$. Then there is $B \lhd A$ such that $\sigma(B) \geq r$.*

*Proof* We seek $x \in {}^*\mathbb{N}$ such that $\sigma(^*A - x) \geq r$. Take $y \in {}^*\mathbb{N}$ such that $d(^*A - y) \geq r$. By the previous lemma, for each $n \in \mathbb{N}$, there is $z_n \in {}^*\mathbb{N}$ with $z_n \geq y$ such that $\sigma(^*A - z_n) \geq r - 1/n$. By overflow, for each $n \in \mathbb{N}$, there is infinite $K_n \in {}^*\mathbb{N}$ such that, for each $m \le K_n$, we have

$$\delta((^*A - z_n), m) \ge r - 1/n.$$

Take an infinite $K \in {}^*\mathbb{N}$ such that $K \le K_n$ for each $n$. (This is possible by countable saturation.) Let

$$D := \{\alpha \in {}^*\mathbb{N} \ : \ (\exists z \in {}^*\mathbb{N})(\forall m \le K)\delta((^*A - z), m) \ge r - 1/\alpha\}.$$

Then $D$ is internal and $\mathbb{N} \subseteq D$, whence by overflow there is infinite $N \in D$. Take $x \in {}^*\mathbb{N}$ such that $\delta((^*A - x), m) \ge r - 1/N$ for all $m \le N$. In particular, for all $m \in \mathbb{N}$, we have $\delta((^*A - x), m) \ge r$, whence this $x$ is as desired.

Theorem 11.11 and Proposition 11.14 immediately yield:

**Corollary 11.15** $\mathrm{BD}(A) \ge r$ if and only if there is $B \lhd A$ such that $\sigma(B) \ge r$.

We end this section with a curious application of Proposition 11.14. We will make more serious use of this technique in the next section.

**Proposition 11.16** *Szemeredi's Theorem is equivalent to the following (apparently weaker statement): There exists $\epsilon > 0$ such that every set $A \subseteq \mathbb{N}$ with $\sigma(A) \ge 1 - \epsilon$ contains arbitrarily long arithmetic progressions.*

*Proof* Fix $A \subseteq \mathbb{N}$ with $\mathrm{BD}(A) > 0$; we wish to show that $A$ contains arbitrarily long arithmetic progressions. By Proposition 10.10, there is $k \in \mathbb{N}$ such that $\mathrm{BD}(A + [0, k]) \ge 1 - \epsilon$. If $A + [0, k]$ contains arbitrarily long arithmetic progressions, then by van der Waerden's theorem, there is $i \in [0, k]$ such that $A + i$ contains arbitrarily long arithmetic progressions, whence so does $A$. It follows that we may assume that $\mathrm{BD}(A) \ge 1 - \epsilon$.

By Proposition 11.14, we have $B \lhd A$ such that $\sigma(B) \ge 1 - \epsilon$, whence, by assumption, we have that $B$ contains arbitrarily long arithmetic progressions, and hence so does $A$.

## 11.3  Applications

We use the ideas from the preceding section to derive some Banach density versions of theorems about Shnirelman density. We first recall the following result of Shnirleman (see, for example, [65, page 8]):

**Theorem 11.17** *Suppose that $A \subseteq \mathbb{N}_0$ is such that $0 \in A$ and $\sigma(A) > 0$. Then $A$ is a basis, that is, there is $h \in \mathbb{N}$ such that $\Sigma_h(A) = \mathbb{N}$.*

Using nonstandard methods, Jin was able to prove a Banach density version of the aforementioned result:

**Theorem 11.18** *Suppose that $A \subseteq \mathbb{N}$ is such that $\gcd(A - \min(A)) = 1$ and $BD(A) > 0$. Then $A$ is a Banach basis, that is, there is $h \in \mathbb{N}$ such that $\Sigma_h(A)$ is thick.*

Note that we must assume that $\gcd(A - \min(A)) = 1$, for if $\gcd(A - \min(A)) = c > 1$, then $hA \subseteq \{h \min(A) + nc \ : \ n \in \mathbb{N}\}$, which does not contain arbitrarily long intervals.

*Proof (of Theorem 11.18)* Suppose $BD(A) = r$ and $\gcd(A - \min(A)) = 1$. The latter property guarantees the existence of $m \in \mathbb{N}$ such that $\Sigma_m(A - \min(A))$ contains two consecutive numbers, whence $c, c + 1 \in \Sigma_m(A)$ for some $c \in \mathbb{N}$. By Proposition 11.14, there is $a \in {}^*\mathbb{N}$ such that $\sigma({}^*A - a + 1) \geq r$. In particular, $a \in {}^*A$. Consequently, we have

$$\sigma(\Sigma_{1+m}({}^*A) - a - c) \geq \sigma({}^*A + \{c, c + 1\} - a - c) \geq \sigma({}^*A - a + 1) \geq r.$$

Since $0 \in \Sigma_{1+m}({}^*A) - a - c$, Shnireleman's theorem implies that there is $n$ such that $\mathbb{N} \subseteq \Sigma_n(\Sigma_{1+m}({}^*A) - a - c)$. By , there is $N$ such that $[0, N] \subseteq \Sigma_n(\Sigma_{1+m}({}^*A) - a - c)$. Set $h := n(1 + m)$, so $[0, N] + n(a + c) \subseteq {}^*(\Sigma_h(A))$. By transfer, $\Sigma_h(A)$ contains arbitrarily long intervals. ∎

With similar methods, one can prove the Banach density analogue of the following theorem of Mann (see, for example, [65, page 5]):

**Theorem 11.19** *Given $A, B \subseteq \mathbb{N}_0$ such that $0 \in A \cap B$, we have $\sigma(A + B) \geq \min\{\sigma(A) + \sigma(B), 1\}$.*

Observe that the exact statement of Mann's theorem is false if one replaces Shnirelman density by Banach density. Indeed, if $A$ and $B$ are both the set of even numbers, then $BD(A + B) = \frac{1}{2}$ but $BD(A) + BD(B) = 1$. However, if one replaces $A + B$ by $A + B + \{0, 1\}$, the Banach density version of Mann's theorem is true.

**Theorem 11.20** *Given $A, B \subseteq \mathbb{N}$, we have $BD(A + B + \{0, 1\}) \geq \min\{BD(A) + BD(B), 1\}$.*

The idea behind the proof of Theorem 11.20 is, as before, to reduce to the case of Shnirelman density by replacing the given sets with hyperfinite shifts. In the course of the proof of Theorem 11.20, we will need to use the following fact from additive number theory (see, for example, [65, page 6]):

**Theorem 11.21 (Besicovitch's Theorem)** *Suppose $A, B \subseteq \mathbb{N}$ and $s \in [0, 1]$ are such that $1 \in A$, $0 \in B$, and $|B \cap [1, n]| \geq s(n + 1)$ for every $n \in \mathbb{N}$. Then $\sigma(A + B) \geq \min\{\sigma(A) + \sigma(B), 1\}$.*

*Proof (of Theorem 11.20)* Set $r := \mathrm{BD}\,(A)$ and $s := \mathrm{BD}\,(B)$. We can assume, without loss of generality, that $r \leq s \leq 1/2$. By Proposition 11.14, one can find $a \in {}^*A$ and $b \in {}^*B$ such that $\sigma\,({}^*A - a + 1) \geq r$ and $\sigma\,({}^*B - b + 1) \geq s$.

*Claim* For every $n \in \mathbb{N}$, one has that $|({}^*B + \{0, 1\}) \cap [b + 1, b + n]| \geq s\,(n + 1)$.

*Proof of Claim* Let $[1, k_0]$ be the largest initial segment of $\mathbb{N}$ contained in $({}^*B + \{0, 1\} - b) \cap \mathbb{N}$ (if no such $k_0$ exists, then the claim is clearly true) and let $[1, k_1]$ be the largest initial segment of $\mathbb{N}$ disjoint from $(({}^*B + \{0, 1\}) - (b + k_0)) \cap \mathbb{N}$. We note the following:

- For $1 \leq n \leq k_0$, we have that

$$\left|({}^*B + \{0, 1\}) \cap [b + 1, b + n]\right| = n \geq (n + 1)/2 \geq s\,(n + 1).$$

- For $k_0 + 1 \leq n < k_0 + k_1$, since $\sigma\,({}^*B - b + 1) \geq s$, we have that

$$
\begin{aligned}
\left|({}^*B + \{0, 1\}) \cap [b + 1, b + n]\right| &\geq \left|{}^*(B + 1) \cap [b + 1, b + n]\right| \\
&= \left|{}^*(B + 1) \cap [b + 1, b + n + 1]\right| \\
&\geq s\,(n + 1).
\end{aligned}
$$

- For $n \geq k_0 + k_1$, since $k_0 + k_1 + 1 \in {}^*B$, $k_0 + k_1 + 1 \notin {}^*B + 1$, and $\sigma\,({}^*B - b + 1) \geq s$, we have that

$$
\begin{aligned}
\left|({}^*B + \{0, 1\}) \cap [b + 1, b + n]\right| &\geq \left|{}^*(B + 1) \cap [b + 1, b + n]\right| + 1 \\
&\geq sn + 1 \geq s\,(n + 1).
\end{aligned}
$$

These observations conclude the proof of the claim.

One can now apply Besicovitch's theorem to ${}^*A - a + 1$ and ${}^*B + \{0, 1\} - b$ (intersected with $\mathbb{N}$) to conclude that

$$\sigma\left(({}^*A - a + 1) + ({}^*B + \{0, 1\} - b)\right) \geq \min\left\{\sigma\left({}^*A - a + 1\right) + s, 1\right\} \geq r + s.$$

Finally, observe that

$$\left({}^*A - a + 1\right) + \left({}^*B + \{0, 1\} - b\right) = {}^*(A + B + \{0, 1\}) - (a + b).$$

Hence

$$\mathrm{BD}\,(A + B + \{0, 1\}) \geq \sigma\left({}^*(A + B + \{0, 1\}) - (a + b)\right) \geq r + s.$$

## Notes and References

The notion of finite embeddability was isolated and studied in [36], although it was implicit in several previous papers of additive number theory. The idea of extending the finite embeddability relation to ultrafilters is due to Luperi Baglini, and it was studied jointly with Blass and Di Nasso in [18, 96]). The material in Sections 11.2 and 11.3 is from Jin's paper [77].

# Chapter 12
# Jin's Sumset Theorem

In this chapter, we state and prove Jin's Sumset Theorem, which is one of the earliest results in combinatorial number theory proven using nonstandard methods. We present Jin's original nonstandard proof, as well as an ultrafilter proof due to Beiglböck and an alternative nonstandard proof due to Di Nasso. In the final section, we prove a recent quantitative strengthening of Jin's Sumset Theorem.

## 12.1 The Statement of Jin's Sumset Theorem and Some Standard Consequences

**Definition 12.1** An initial segment U of $^*\mathbb{N}_0$ is a *cut* if $U + U \subseteq U$.

**Exercise 12.2** If U is a cut, then either U is external or else $U = {}^*\mathbb{N}$.

*Example 12.3*

1. $\mathbb{N}$ is a cut.
2. If $N$ is an infinite element of $^*\mathbb{N}$, then $U_N := \{x \in {}^*\mathbb{N} : \frac{x}{N} \approx 0\}$ is a cut.

Fix a cut U of $^*\mathbb{N}$ and suppose that $U \subseteq [0, N)$. Given $x, y \in {}^*\mathbb{N}$, we write $x \sim_U y$ if $|x - y| \in U$; note that $\sim_U$ is an equivalence relation on $^*\mathbb{N}$. We let $[x]_{U,N}$, or simply $[x]_N$ if no confusion can arise, denote the equivalence class of $x$ under $\sim_U$ and we let $[0, N)/U$ denote the set of equivalence classes. We let $\pi_U : [0, N) \to [0, N)/U$ denote the quotient map. The linear order on $[0, N)$ descends to a linear order on $[0, N)/U$. Moreover, one can push forward the Loeb measure on $[0, N)$ to a measure on $[0, N)/U$, which we also refer to as Loeb measure.

© Springer Nature Switzerland AG 2019
M. Di Nasso et al., *Nonstandard Methods in Ramsey Theory and Combinatorial Number Theory*, Lecture Notes in Mathematics 2239,
https://doi.org/10.1007/978-3-030-17956-4_12

*Example 12.4* Fix $N \in {}^*\mathbb{N}$ infinite and consider the cut $\mathrm{U}_N$ from Example 12.3. Note that the surjection $f : [0, N) \to [0, 1]$ given by $f(\beta) := \mathrm{st}(\beta/N)$ descends to a bijection of ordered sets $f : [0, N)/\mathrm{U}_N \to [0, 1]$. The discussion in Sect. 5.3 of Chap. 5 shows that the measure on $[0, 1]$ induced by the Loeb measure on $[0, N)/\mathrm{U}_N$ via $f$ is precisely Lebesgue measure.

For any cut $\mathrm{U}$ contained in $[0, N)$, the set $[0, N)/\mathrm{U}$ has a natural topology induced from the linear order, whence it makes sense to talk about category notions in $[0, N)/\mathrm{U}$. (This was first considered in [84].) It will be convenient to translate the category notions from $[0, N)/\mathrm{U}$ back to $[0, N]$:

**Definition 12.5** $A \subseteq [0, N)$ is $\mathrm{U}$-*nowhere dense* if $\pi_{\mathrm{U}}(A)$ is nowhere dense in $[0, N)/U$. More concretely: $A$ is $\mathrm{U}$-nowhere dense if, given any $a < b$ in $[0, N)$ with $b - a > \mathrm{U}$, there is $[c, d] \subseteq [a, b]$ with $d - c > \mathrm{U}$ such that $[c, d] \subseteq [0, N)\backslash A$. If $A$ is not $\mathrm{U}$-nowhere dense, we say that $A$ is $\mathrm{U}$-*somewhere dense*.

Recall the following famous theorem of Steinhaus:

**Theorem 12.6** *If* $C, D \subseteq [0, 1]$ *have positive Lebesgue measure, then* $C + D$ *contains an interval.*

For $x, y \in [0, N)$, set $x \oplus_N y := x + y \mod N$. For $A, B \subseteq [0, N)$, set

$$A \oplus_N B := \{x \oplus_N y : x \in A, y \in B\}.$$

In light of Example 12.4, Theorem 12.6 says that whenever $A', B \subseteq [0, N)$ are internal sets of positive Loeb measure, then $A \oplus_N B$ is $\mathrm{U}_N$-somewhere dense. Keisler and Leth asked whether or not this is the case for any cut. Jin answered this positively in [78]:

**Theorem 12.7 (Jin's Sumset Theorem)** *If* $\mathrm{U} \subseteq [0, N)$ *is a cut and* $A, B \subseteq [0, N)$ *are internal sets with positive Loeb measure, then* $A \oplus_N B$ *is* $\mathrm{U}$-*somewhere dense.*

**Exercise 12.8** Prove Theorem 12.6 from Theorem 12.7.

We will prove Theorem 12.7 in the next section. We now prove the following standard corollary of Theorem 12.7, which is often also referred to as Jin's sumset theorem, although this consequence was known to Leth beforehand.

**Corollary 12.9** *Suppose that* $A, B \subseteq \mathbb{N}$ *have positive Banach density. Then* $A + B$ *is piecewise syndetic.*

*Proof* Set $r := \mathrm{BD}(A)$ and $s := \mathrm{BD}(B)$. Fix $N \in {}^*\mathbb{N}$ infinite and take $x, y \in {}^*\mathbb{N}$ such that

$$\delta({}^*A \cap [x, x + N)|) \approx r, \quad \delta({}^*B \cap [y, y + N)) \approx s.$$

Let $C := {}^*A - x$ and $D := {}^*B - y$, so we may view $C$ and $D$ as internal subsets of $[0, 2N)$ of positive Loeb measure. By Jin's theorem applied to the cut $\mathbb{N}$, we have that $C \oplus_{2N} D = C + D$ is $\mathbb{N}$-somewhere dense, that is, there is a hyperfinite interval

$I$ such that all gaps of $C + D$ on $I$ have finite length. By , there is $m \in \mathbb{N}$ such that all gaps of $C + D$ on $I$ have length at most $m$. Therefore, $x + y + I \subseteq {}^*(A + B + [0, m])$. By transfer, for any $k \in \mathbb{N}$, $A + B + [0, m]$ contains an interval of length $k$, whence $A + B$ is piecewise syndetic.

It is interesting to compare the previous corollary to Proposition 10.25. It is also interesting to point out that Corollary 12.9 can also be used to give an alternative proof of Theorem 11.18. Indeed, suppose $BD(A) > 0$ and $\gcd(A - \min(A)) = 1$. Then there is $h \in \mathbb{N}$ such that $A + A + [0, h]$ is thick. It follows that $A + A + [x, x + h]$ is thick for all $x \in \mathbb{N}$. As in the proof of Theorem 11.18, take $m$ and consecutive $a, a + 1 \in \Sigma_m(A)$. Note that, for all $i = 0, 1, \ldots, h$, we have that $ha + i = i(a + 1) + (h - i)a \in \Sigma_{hm}(A)$. It follows that $A + A + [ha, ha + h] \subseteq \Sigma_{hm+2}(A)$, whence $\Sigma_{hm+2}(A)$ is thick.

## 12.2   Jin's Proof of the Sumset Theorem

We now turn to the proof of Theorem 12.7 given in [78]. Suppose, towards a contradiction, that there is a cut U for which the theorem is false. If $H > U$ and $A, B \subseteq [0, H)$ are internal, we say that $(A, B)$ is $(H, U)$-*bad* if $\mu_H(A), \mu_H(B) > 0$ and $A \oplus_H B$ is U-nowhere dense. We set

$$r := \sup\{\mu_H(A) \ : \ (A, B) \text{ is } (H, U) \text{ bad for some } H > U \text{ and some } B \subseteq [0, H)\}.$$

By assumption, $r > 0$. We fix $\epsilon > 0$ sufficiently small. We then set

$$s := \sup\{\mu_H(B) \ : \ (A, B) \text{ is } (H, U)\text{-bad for some } H > U$$
$$\text{and some } A \subseteq [0, H) \text{ with } \mu_H(A) > r - \epsilon\}.$$

By the definition of $r$, we have that $s > 0$. Also, by the symmetry of the definition of $r$, we have that $r \geq s$. The following is slightly less obvious:

*Claim 1*  $s < \frac{1}{2} + \epsilon$.

*Proof of Claim 1*  Suppose, towards a contradiction, that $s \geq \frac{1}{2} + \epsilon$. We may thus find $H > \mathbb{N}$ and an $(H, U)$-bad pair $(A, B)$ with $\mu_H(A) > \frac{1}{2}$ and $\mu_H(B) > \frac{1}{2}$. Since addition modulo $H$ is translation invariant, it follows that for any $x \in [0, H)$, we have that $A \cap (x \ominus_H B) \neq \emptyset$, whence $A \oplus_H B = [0, H - 1)$, which is a serious contradiction to the fact that $A \oplus_H B$ is U-nowhere dense.

We now fix $\delta > 0$ sufficiently small, $H > U$ and an $(H, U)$-bad $(A, B)$ such that $\mu_H(A) > r - \epsilon$ and $\mu_H(B) > s - \delta$. We will obtain a contradiction by producing $K > U$ and $(K, U)$-bad $(A', B')$ such that $\mu_K(A') > r - \epsilon$ and $\mu_K(B') > s + \delta$, contradicting the definition of $s$.

We first show that it suffices to find $K > U$ such that $K/H \approx 0$ and such that there are hyperfinite intervals $I, J \subseteq [0, H)$ of length $K$ for which

$$\mathrm{st}\left(\frac{|A \cap I|}{K}\right) > r - \epsilon \text{ and } \mathrm{st}\left(\frac{|B \cap J|}{K}\right) > s + \delta.$$

Indeed, suppose that $I := [a, a + K)$ and $J := [b, b + K)$ are as above. Let $A' := (A \cap I) - a$ and $B' := (B \cap J) - b$. Then $\mu_K(A') > r - \epsilon$ and $\mu_K(B') > s + \delta$. It remains to see that $(A', B')$ is $(K, U)$-bad. Since $A \oplus_H B$ is U-nowhere dense, it is clear that $(A \cap I) \oplus_H (B \cap J)$ is also U-nowhere dense. Since $A' \oplus_H B' = ((A \cap I) \oplus_H (B \cap J)) \ominus (a + b)$, we have that $A' \oplus_H B'$ is U-nowhere dense. Since $K/H$ is infinitesimal, we have that $A' \oplus_H B' = A' \oplus_{2K} B'$. It follows that $A' \oplus_K B'$ is the union of two U-nowhere dense subsets of $[0, K)$, whence is also U-nowhere dense, and thus $(A', B')$ is $(K, U)$-bad, as desired.

We now work towards finding the appropriate $K$. By the definition of U-nowhere dense, we have, for every $k \in U$, that $A \oplus_H (B \oplus_H [-k, k])) = (A \oplus_H B) \oplus_H [-k, k]$ is U-nowhere dense. By the definition of $s$, it follows that $\mu_H(B \oplus_H [-k, k]) \leq s$ for each $k \in U$. Since $U$ is external and closed under addition, it follows that there is $K > U$ with $K/H$ infinitesimal such that

$$\frac{|B \oplus_H [-K, K]|}{H} \leq s + \frac{\delta}{2}.$$

We finish by showing that this $K$ is as desired.

Let $\mathscr{I} := \{[iK, (i+1)K) : 0 \leq i \leq H/K - 1\}$ be a partition of $[0, H-1)$ into intervals of length $K$ (with a negligible tail omitted). Let $X := \{i \in [0, H/K - 1] : [iK, (i+1)K - 1) \cap B = \emptyset\}$.

*Claim 2* $\frac{|X|}{|\mathscr{I}|} > \frac{1}{3}$.

*Proof of Claim 2* Suppose, towards a contradiction, that $\frac{|X|}{|\mathscr{I}|} \leq \frac{1}{3}$. Fix $i \notin X$ and $x \in [iK, (i+1)K)$. Write $x = iK + j$ with $j \in [0, K - 1]$. Since $i \notin X$, there is $l \in [0, K - 1)$ such that $iK + l \in B$. It follows that $x = (iK + l) + (j - l) \in B \oplus_H [-K, K]$. Consequently,

$$|B \oplus_H [-K, K]| \geq \sum_{i \notin X} K \geq \frac{2}{3}(H/K - 1) \cdot K = \frac{2}{3}H - \frac{2}{3}K,$$

whence

$$\frac{|B \oplus_H [-K, K]|}{H} \geq \frac{2}{3} - \frac{2}{3}\frac{K}{H} \approx \frac{2}{3},$$

which, for sufficiently small $\epsilon$ and $\delta$, contradicts the fact that $\frac{|B \oplus_H [-K, K]|}{H} \leq s + \frac{\delta}{2}$.

Let $\mathscr{I}' := \{[iK, (i+1)K) : i \notin X\}$. As explained above, the following claim completes the proof of the theorem.

*Claim 3* There are $I, J \in \mathscr{I}$ such that

$$\mathrm{st}\left(\frac{|A \cap I|}{K}\right) > r - \epsilon \text{ and } \mathrm{st}\left(\frac{|B \cap J|}{K}\right) > s + \delta.$$

*Proof of Claim 3* We only prove the existence of $J$; the proof of the existence of $I$ is similar (and easier). Suppose, towards a contradiction, that $\mathrm{st}(\frac{|B \cap J|}{K}) \leq s + \delta$ for all $J \in \mathscr{I}$. We then have

$$s - \delta < \frac{|B \cap [0, H-1)|}{H} = \frac{1}{H} \sum_{J \in \mathscr{I}'} |B \cap [iK, (i+1)K)|$$

$$\leq \frac{1}{H} \cdot \frac{2}{3} \cdot (H/K) \cdot (s + \delta)K = \frac{2}{3}(s + \delta).$$

If $\delta \leq \frac{s}{5}$, then this yields a contradiction.

## 12.3 Beiglböck's Proof

It is straightforward to verify that Corollary 12.9 is also true for subsets of $\mathbb{Z}$:

**Corollary 12.10** *If $A, B \subseteq \mathbb{Z}$ are such that $\mathrm{BD}(A), \mathrm{BD}(B) > 0$, then $A + B$ is piecewise syndetic.*

In this section, we give Beiglböck's ultrafilter proof of Corollary 12.10 appearing in [7]. We first start with some preliminary facts on invariant means on $\mathbb{Z}$.

**Definition 12.11** An *invariant mean* on $\mathbb{Z}$ is a linear functional $\ell : B(\mathbb{Z}) \to \mathbb{R}$ that satisfies the following properties:

1. $\ell$ is positive, that is, $\ell(f) \geq 0$ if $f \geq 0$;
2. $\ell(1) = 1$; and
3. $\ell(k.f) = \ell(f)$ for all $k \in \mathbb{Z}$ and $f \in B(\mathbb{Z})$, where $(k.f)(x) := f(x - k)$.

There are many invariant means on $\mathbb{Z}$:

**Exercise 12.12** Suppose that $(I_n)$ is a sequence of intervals in $\mathbb{Z}$ with $|I_n| \to \infty$ as $n \to \infty$. Fix $\mathscr{U} \in \beta\mathbb{N}$. Define, for $f \in B(\mathbb{Z})$, $\ell(f) = \lim_{\mathscr{U}} (\frac{1}{|I_n|} \sum_{x \in I_n} f(x))$. Show that $\ell$ is an invariant mean on $\mathbb{Z}$.

In fact, we have:

**Lemma 12.13** *For every $A \subseteq \mathbb{Z}$, there is an invariant mean $\ell$ on $\mathbb{Z}$ such that $\ell(1_A) = \mathrm{BD}(A)$.*

*Proof* Let $(I_n)$ be a sequence of intervals witnessing the Banach density of $A$. Fix nonprincipal $\mathcal{U} \in \beta\mathbb{Z}$. Define $\ell$ as in Exercise 12.12 for these choices of $(I_n)$ and $\mathcal{U}$. It is clear that $\ell(1_A) = \mathrm{BD}(A)$.

**Lemma 12.14** *For every invariant mean $\ell$ on $\mathbb{Z}$, there is a regular Borel probability measure $v$ on $\beta\mathbb{Z}$ such that $\ell(1_A) = v(\overline{A})$ for every $A \subseteq \mathbb{Z}$.*

*Proof* Fix a mean $\ell$ on $\mathbb{Z}$. Since $f \mapsto \beta f$ yields an isomorphism $B(\mathbb{Z}) \cong C(\beta\mathbb{Z})$, the Riesz Representation Theorem yields a regular Borel probability measure $v$ on $\beta\mathbb{Z}$ such that $\ell(f) = \int_{\beta\mathbb{Z}} (\beta f) dv$ for all $f \in B(\mathbb{Z})$. In particular,

$$\ell(1_A) = \int_{\beta\mathbb{Z}} (\beta 1_A) dv = v(\overline{A}).$$

The following lemma is the key to Beiglböck's proof of Corollary 12.10.

**Lemma 12.15** *For any $A, B \subseteq \mathbb{Z}$, there is $\mathcal{U} \in \beta\mathbb{Z}$ such that $\mathrm{BD}(A \cap (B - \mathcal{U})) \geq \mathrm{BD}(A) \cdot \mathrm{BD}(B)$.*

*Proof* Fix an invariant mean $\ell$ on $\mathbb{Z}$ such that $\ell(1_B) = \mathrm{BD}(B)$ and let $v$ be the associated Borel probability measure on $\beta\mathbb{Z}$. Let $(I_n)$ be a sequence of intervals witnessing the Banach density of $A$. Define $f_n : \beta\mathbb{Z} \to [0, 1]$ by

$$f_n(\mathcal{U}) := \delta((A \cap (B - \mathcal{U}), I_n) = \frac{1}{|I_n|} \sum_{k \in A \cap I_n} 1_{\overline{B-k}}(\mathcal{U}).$$

Set $f(\mathcal{U}) := \limsup_n f_n(\mathcal{U})$ and note that $f(\mathcal{U}) \leq \mathrm{BD}(A \cap (B - \mathcal{U}))$ for all $\mathcal{U} \in \beta\mathbb{Z}$. Fatou's Lemma implies

$$\int_{\beta\mathbb{Z}} f dv \geq \limsup_n \int_{\beta\mathbb{Z}} \frac{1}{|I_n|} \sum_{k \in A \cap I_n} 1_{U_{B-k}} dv = \limsup_n \frac{1}{|I_n|} \sum_{k \in I_n \cap A} \ell(1_{B-k}).$$

Since $\ell$ is invariant, the latter term is equal to $\limsup_n \delta(A, I_n) \cdot \ell(1_B) = \mathrm{BD}(A) \cdot \mathrm{BD}(B)$. Thus, we have shown $\int_{\beta\mathbb{Z}} f dv \geq \mathrm{BD}(A) \cdot \mathrm{BD}(B)$. In particular, there is some $\mathcal{U} \in \mathbb{Z}$ such that $f(\mathcal{U}) \geq \mathrm{BD}(A) \cdot \mathrm{BD}(B)$, as desired.

Notice that, in the notation of the above proof, $\mu(\mathbb{Z}) = 0$, whence we can take $\mathcal{U}$ as in the conclusion of the lemma to be nonprincipal.

We can now give Beiglböck's proof of Corollary 12.10. Assume that $\mathrm{BD}(A), \mathrm{BD}(B) > 0$. Apply the previous lemma with $A$ replaced by $-A$ (which has the same Banach density), obtaining $\mathcal{U} \in \beta\mathbb{Z}$ such that $C := (-A) \cap (B - \mathcal{U})$ has positive Banach density. By Lemma 10.25, $C - C$ is syndetic; since $C - C \subseteq A + (B - \mathcal{U})$, we have that $A + (B - \mathcal{U})$ is also syndetic.

Suppose $s \in A + (B - \mathcal{U})$. Then for some $a \in A$, $B - (s - a) \in \mathcal{U}$, whence $a + B - s \in \mathcal{U}$ and hence $A + B - s \in \mathcal{U}$. Thus, for any finite set $s_1, \ldots, s_n \in A + (B - \mathcal{U})$, we have $\bigcap_{i=1}^n (A + B - s_i) \in \mathcal{U}$, and, in particular, is nonempty, meaning there is $t \in \mathbb{Z}$ such that $t + \{s_1, \ldots, s_n\} \subseteq A + B$. We claim that this implies that $A + B$ is piecewise syndetic. Indeed, take $F \subseteq \mathbb{Z}$ such that $F + A + (B - \mathcal{U}) = \mathbb{Z}$. We claim that $F + A + B$ contains arbitrarily long intervals. To see this, fix $n \in \mathbb{N}$ and, for $i = 1, \ldots, n$ take $s_i \in A + (B - \mathcal{U})$ such that $i \in F + s_i$. Take $t \in \mathbb{Z}$ such that $t + \{s_1, \ldots, s_n\} \subseteq A + B$. Then $t + [1, n] \subseteq t + F + \{s_1, \ldots, s_n\} \subseteq F + (A + B)$, completing the proof.

## 12.4   A Proof with an Explicit Bound

A proof of Corollary 12.10 can be given by using a simple counting argument of finite combinatorics in the nonstandard setting. In this way, one also obtains an explicit bound on the number of shifts of the sumset that are needed to produce a thick set.

**Lemma 12.16** *Let $C \subseteq [1, n]$ and $D \subseteq [1, m]$ be finite sets of natural numbers. Then there exists $k \leq n$ such that*

$$\frac{|(C - k) \cap D|}{m} \geq \frac{|C|}{n} \cdot \frac{|D|}{m} - \frac{|D|}{n}.$$

*Proof* If $\chi : [1, n] \to \{0, 1\}$ is the characteristic function of $C$, then for every $d \in D$, we have

$$\frac{1}{n} \cdot \sum_{k=1}^{n} \chi(k + d) = \frac{|C \cap [1 + d, n + d]|}{n} = \frac{|C|}{n} + \frac{e(d)}{n}$$

where $|e(d)| \leq d$. Then:

$$\frac{1}{n} \cdot \sum_{k=1}^{n} \left( \frac{1}{m} \cdot \sum_{d \in D} \chi(k + d) \right) = \frac{1}{m} \cdot \sum_{d \in D} \left( \frac{1}{n} \cdot \sum_{x=1}^{n} \chi(k + d) \right)$$

$$= \frac{1}{m} \cdot \sum_{d \in D} \frac{|C|}{n} + \frac{1}{nm} \cdot \sum_{d \in D} e(d) = \frac{|C|}{n} \cdot \frac{|D|}{m} + e$$

where

$$|e| = \left| \frac{1}{nm} \sum_{d \in D} e(d) \right| \leq \frac{1}{nm} \sum_{d \in D} |e(d)| \leq \frac{1}{nm} \cdot \sum_{d \in D} d \leq \frac{1}{nm} \sum_{d \in D} m = \frac{|D|}{n}.$$

By the *pigeonhole principle*, there must exist at least one number $k \leq n$ such that

$$\frac{|(C - k) \cap D|}{m} = \frac{|(D + k) \cap C|}{m} = \frac{1}{m} \cdot \sum_{d \in D} \chi(k + d) \geq \frac{|C|}{n} \cdot \frac{|D|}{m} - \frac{|D|}{n}.$$

**Theorem 12.17** *Let* $A, B \subseteq \mathbb{Z}$ *have positive Banach densities* $\mathrm{BD}(A) = \alpha > 0$ *and* $\mathrm{BD}(B) = \beta > 0$. *Then there exists a finite set* $F$ *with* $|F| \leq \frac{1}{\alpha\beta}$ *such that* $(A + B) + F$ *is thick. In particular,* $A + B$ *is piecewise syndetic.*

*Proof* Pick infinite $v, N \in {}^*\mathbb{N}$ such that $v/N \approx 0$, and pick intervals $[\Omega + 1, \Omega + N]$ and $[\Xi + 1, \Xi + v]$ such that

$$\frac{|{}^*A \cap [\Omega + 1, \Omega + N]|}{N} \approx \alpha \quad \text{and} \quad \frac{|(-{}^*B) \cap [\Xi + 1, \Xi + v]|}{v} \approx \beta.$$

By applying the nonstandard version of the previous lemma to the hyperfinite sets $C = ({}^*A - \Omega) \cap [1, N] \subseteq [1, N]$ and $D = (-{}^*B - \Xi) \cap [1, v]$, one obtains the existence of a number $\zeta$ such that

$$\frac{|(C - \zeta) \cap D|}{v} \geq \frac{|C|}{N} \cdot \frac{|D|}{v} - \frac{|D|}{N} \approx \alpha\beta.$$

Finally, apply Lemma 10.26 to the internal set $E = (C - \zeta) \cap D \subseteq [1, v]$. Since $|E|/v \approx \alpha\beta$, there exists a finite $F \subset \mathbb{Z}$ with $|F| \leq \frac{1}{\alpha\beta}$ and such that $\mathbb{Z} \subseteq (E - E) + F$, and hence, by *overflow*, $I \subseteq (E - E) + F$ for some infinite interval $I$. Since $E \subseteq {}^*A - \Omega$ and $E \subset -{}^*B - \Xi$, it follows that ${}^*(A + B + F) = {}^*A + {}^*B + F$ includes the infinite interval $I + \Omega + \Xi + \zeta$, and hence it is thick.

## 12.5   Quantitative Strengthenings

We end this chapter by proving some technical strengthenings of Corollary 12.10. Indeed, in light of Lemma 10.23, the following theorem can be viewed as a "quantitative" strengthening of Corollary 12.10:

**Theorem 12.18** *Suppose that* $(I_n)$ *is a sequence of intervals with* $|I_n| \to \infty$ *as* $n \to \infty$. *Suppose that* $A, B \subseteq \mathbb{Z}$ *and* $\mathrm{BD}(B) > 0$. *Then:*

1. *If* $\overline{d}_{(I_n)}(A) \geq r$, *then there is a finite set* $F \subseteq \mathbb{Z}$ *such that, for every finite set* $L \subseteq \mathbb{Z}$, *we have*

$$\overline{d}_{(I_n)}\left(\bigcap_{x \in L}(A + B + F + x)\right) \geq r.$$

2. *If* $\underline{d}_{(I_n)}(A) \geq r$, *then for every* $\epsilon > 0$, *there is a finite set* $F \subseteq \mathbb{Z}$ *such that, for every finite set* $L \subseteq \mathbb{Z}$, *we have*

$$\underline{d}_{(I_n)}\left(\bigcap_{x \in L}(A + B + F + x)\right) \geq r - \epsilon.$$

In connection with item (2) of the previous theorem, it will turn out that $F$ depends only on $B$ and $\epsilon$ (but not on $A$ or $(I_n)$). Moreover, item (2) is false if $r - \epsilon$ is replaced by $r$; see [43].

In order to prove Theorem 12.18, we need a preparatory counting lemma.

**Lemma 12.19** *Suppose that $(I_n)$ is a sequence of intervals in $\mathbb{Z}$ such that $|I_n| \to \infty$ as $n \to \infty$. Further suppose $I$ is an infinite hyperfinite interval in $^*\mathbb{Z}$ and $A \subseteq \mathbb{Z}$.*

*1. If $\overline{d}_{(I_n)}(A) \geq r$, then there is $N > \mathbb{N}$ such that*

$$\delta(^*A, I_N) \gtrsim r \quad and \quad \frac{1}{|I_N|} \sum_{x \in I_N} \delta(x - (^*A \cap I_N), I) \gtrsim r. \quad (\dagger)$$

*2. If $\underline{d}_{(I_n)}(A) > r$, then there is $N_0 > \mathbb{N}$ such that $(\dagger)$ holds for all $N \geq N_0$.*

*Proof* For (1), first apply transfer to the statement "for every finite interval $J \subseteq \mathbb{Z}$ and every natural number $k$, there exists $n \geq k$ such that

$$\delta(A, I_n) > r - 2^{-k} \quad and \quad \frac{1}{|I_n|} \sum_{x \in J} |(I_n - x) \triangle I_n| < 2^{-k}."$$

Fix $K > \mathbb{N}$ and let $N$ be the result of applying the transferred statement to $I$ and $K$. Set $C = {}^*A \cap I_N$ and let $\chi_C$ denote the characteristic function of $C$. We have

$$\frac{1}{|I_N|} \sum_{x \in I_N} \delta((x - C), Y) = \frac{1}{|I_N|} \sum_{x \in I_N} \frac{1}{|I|} \sum_{y \in I} \chi_C(x - y)$$

$$= \frac{1}{|I|} \sum_{y \in I} \frac{|C \cap (I_N - y)|}{|I_N|}$$

$$\geq \frac{|C|}{|I_N|} - \sum_{y \in I} \frac{|(I_N - y) \triangle I_N|}{|I_N|}$$

$$\approx r.$$

For (2), apply transfer to the statement "for every finite interval $J \subseteq \mathbb{Z}$ and every natural number $k$, there exists $n_0 \geq k$ such that, for all $n \geq n_0$,

$$\delta(A, I_n) > r - 2^{-n_0} \quad and \quad \frac{1}{|I_n|} \sum_{x \in J} |(I_n - x) \triangle I_n| < 2^{-n_0}."$$

Once again, fix $K > \mathbb{N}$ and let $N_0$ be the result of applying the transferred statement to $I$ and $K$. As above, this $N_0$ is as desired.

*Proof (of Theorem 12.18)* Fix an infinite hyperfinite interval $I$ that is good for $B$. (See Proposition 10.28.)

For (1), assume that $\overline{d}_{(I_n)}(A) \geq r$. Let $N$ be as in part (1) of Lemma 12.19 applied to $I$ and $A$. Once again, set $C := {}^*A \cap I_N$. Consider the $\mu_{I_N}$-measurable function

$$f(x) = \mathrm{st}(\delta(x - C, I)).$$

By Lemma 5.18, we have that

$$\int_{I_N} f d\mu_{I_N} = \mathrm{st}\left(\frac{1}{|I_N|} \sum_{x \in I_N} \delta(x - C, I)\right) \geq r,$$

whence there is some standard $s > 0$ such that $\mu_{I_N}(\{x \in I_N : f(x) \geq 2s\}) \geq r$. Setting $\Gamma = \{x \in I_N : \delta(x - C, I) \geq s\}$, we have that $\mu_{I_N}(\Gamma) \geq r$. Since $I$ is good for $B$, we may take a finite subset $F$ of $\mathbb{Z}$ such that

$$\delta({}^*(B + F), I) > 1 - \frac{s}{2}.$$

Fix $x \in \mathbb{Z}$. Since $I$ is infinite, we have that

$$\delta({}^*(B + F + x), I) = \delta({}^*(B + F), (I - x)) \approx \delta({}^*(B + F), I),$$

whence $\delta({}^*(B + F + x), I) > 1 - s$. Thus, for any $y \in \Gamma$, we have that $(y - C) \cap {}^*(B + F + x) \neq \emptyset$. In particular, if $L$ is a finite subset of $\mathbb{Z}$, then $\Gamma \subseteq {}^*\left(\bigcap_{x \in L} A + B + F + x\right)$. Therefore

$$\overline{d}_{(I_n)}\left(\bigcap_{x \in L} A + B + F + x\right) \geq \mu_{I_N}\left({}^*(\bigcap_{x \in L} A + B + F + x)\right) \geq \mu_{I_N}(\Gamma) \geq r.$$

This establishes (1).

Towards (2), note that we may suppose that $\underline{d}_{(I_n)}(A) > r$. Fix $N_0 > \mathbb{N}$ as in part (2) of Lemma 12.19 applied to $I$ and $A$. Fix $N \geq N_0$ and standard $\varepsilon > 0$ with $\varepsilon < r$. Set

$$\Lambda := \{x \in I_N : \delta((x - C), I) \geq \varepsilon\}$$

and observe that $\frac{|\Lambda|}{|I_N|} > r - \varepsilon$. Since $I$ is good for $B$, we may fix a finite subset $F$ of $\mathbb{Z}$ such that

$$\delta({}^*(B + F), I) > 1 - \frac{\varepsilon}{2}.$$

Fix $x \in \mathbb{Z}$. Since $I$ is infinite, arguing as in the proof of part (1), we conclude that

$$\delta^*(B + F + x), I) > 1 - \varepsilon.$$

Fix $L \subseteq \mathbb{Z}$ finite. As in the proof of part (1), it follows that $\Lambda \subseteq {}^*\left(\bigcap_{x \in L} A + B + F + x\right)$ whence

$$\delta\left({}^*\left(\bigcap_{x \in L} A + B + F + x\right), I_N\right) \geq \frac{|\Lambda|}{|I_N|} > r - \varepsilon.$$

Since the previous inequality held for every $N \geq N_0$, by transfer we can conclude that there is $n_0$ such that, for all $n \geq n_0$, we have

$$\delta\left(\left(\bigcap_{x \in L} A + B + F + x\right), I_n\right) \geq r - \epsilon,$$

whence it follows that

$$\underline{d}_{(I_n)}\left(\bigcap_{x \in L} A + B + F + x\right) \geq r - \varepsilon.$$

## Notes and References

The space of cuts was first studied in the paper [78] and Jin's Sumset Theorem solved Problem 9.13 in that paper negatively. Jin gives a purely standard, finitary version of his proof of the Sumset Theorem in [79]; a simplified elementary standard proof was then given in [35]. The proof given in Sect. 12.4 is due to Di Nasso [36]. The original proof of Theorem 12.18 given in [41] used a Lebesgue Density Theorem for the cut spaces $[0, H]/U$. Indeed, one can give a nice proof of Theorem 12.6 using the standard Lebesgue density theorem and Example 12.4 suggested that perhaps a general Lebesgue density theorem holds for cut spaces. Once this was established, the fact that one has many density points was used to strengthen the sumset theorem in the above manner. The proof given in this chapter follows [43], which actually works for all countable amenable groups rather than just $\mathbb{Z}$; other than the fact that Proposition 10.10 is more difficult to prove for amenable groups than it is for $\mathbb{Z}$, there is not much added difficulty in generalizing to the amenable situation. We should also mention that the amenable group version of Corollary 12.10 was first proven by Beiglböck et al. in [8].

# Chapter 13
# Sumset Configurations in Sets of Positive Density

In this section, we discuss a conjecture of Erdős, which states that a set of natural numbers of positive lower density contains the sum of two infinite sets. We begin with the history of the conjecture and discuss its nonstandard reformulation. We then present a proof of the conjecture in the "high density" case, which follows from a "1-shift" version of the conjecture in the general case. We conclude with a discussion of how these techniques yield a weak density version of Folkman's theorem.

## 13.1  Erdős' Conjecture

Just as Szemeredi's theorem is a "density" version of van der Waerden's theorem, it is natural to wonder if the density version of Hindman's theorem is true, namely: does every set of positive density contain an FS set? It is clear that the answer to this question is: no! Indeed, the set of odd numbers has positive density, but does not even contain $PS(B)$ for any infinite set $B$. Here, $PS(B) := \{b + b' : b, b' \in B, b \neq b'\}$. This example is easily fixed if we allow ourselves to translate the original set, so Erdős conjectured that this was the only obstruction to a weak density version of Hindman's theorem, namely: if $A \subseteq \mathbb{N}$ has positive density, then there is $t \in \mathbb{N}$ and infinite $B \subseteq A$ such that $t + FS(B) \subseteq A$. Strauss provided a counterexample to this conjecture[1] (see [45]), whence Erdős changed his conjecture to the following, which we often refer to as Erdős' sumset conjecture (see [103] and [46, p. 85]):

---

[1] It still seems to be open whether or not a set of positive density (of any kind) must contain a translate of $PS(B)$ for some infinite $B$.

© Springer Nature Switzerland AG 2019
M. Di Nasso et al., *Nonstandard Methods in Ramsey Theory and Combinatorial Number Theory*, Lecture Notes in Mathematics 2239, https://doi.org/10.1007/978-3-030-17956-4_13

*Conjecture 13.1* Suppose that $A \subseteq \mathbb{N}$ is such that $\underline{d}(A) > 0$. Then there exist infinite sets $B$ and $C$ such that $B + C \subseteq A$.

It will be convenient to give a name to sets satisfying the conclusion of Conjecture 13.1.

**Definition 13.2** We say that $A \subseteq \mathbb{N}$ has the *sumset property* if there are infinite sets $B, C \subseteq \mathbb{N}$ such that $B + C \subseteq A$.

Many sets that are structurally large have the sumset property as indicated by the following proposition. While this result follows from standard results in the literature, we prefer to give the following elegant argument of Leth.[2]

**Proposition 13.3** *If $A$ is piecewise syndetic, then $A$ has the sumset property. More precisely, there is an infinite set $B \subseteq \mathbb{N}$ and $k \in \mathbb{N}$ such that $\mathrm{PS}(B) - k \subseteq A$.*

*Proof* Since $A$ is piecewise syndetic, there exists $m$ and an interval $[a, b]$ in $^*\mathbb{N}$ with $a$ and $b - a$ infinite such that $^*A$ has no gaps of size larger than $m$ on $[a, b]$. Set $L := (^*A - a) \cap \mathbb{N}$, so that $a + L \subseteq {}^*A$. Let $l$ be the first element in $^*L$ greater than or equal to $a$. Set $k := l - a$. Since $L$ contains no gaps of size larger than $m$, we know that $0 \le k \le m$. We now have:

$$l - k + L \subseteq {}^*A \text{ and } l \in {}^*L.$$

Take $b_0 \in L$ arbitrary. Assume now that $b_0 < b_1 < \cdots < b_n \in L$ have been chosen so that $b_i + b_j - k \in A$ for $1 \le i < j \le n$. Since the statement "there is $l \in {}^*L$ such that $l > b_n$ and $l - k + b_i \in {}^*A$ for $i = 1, \ldots, n$" is true, by transfer there is $b_{n+1} \in L$ such that $b_{n+1} > b_n$ and $b_i + b_{n-1} - k \in A$ for $i = 1, \ldots, n$. The set $B := \{b_0, b_1, b_2, \ldots\}$ defined this way is as desired.

The first progress on Erdős' conjecture was made by Nathanson in [103], where he proved the following:

**Theorem 13.4 (Nathanson)** *If $\mathrm{BD}(A) > 0$ and $n \in \mathbb{N}$, then there are $B, C$ with $\mathrm{BD}(B) > 0$ and $|C| = n$ such that $B + C \subseteq A$.*

This theorem follows immediately by induction using the following lemma. We take the opportunity here to give a short nonstandard proof.

**Lemma 13.5 (Kazhdan)** *Suppose that $\mathrm{BD}(A) > 0$ and $t \in \mathbb{N}$. Then there is $B \subseteq A$ with $\mathrm{BD}(B) > 0$ and $c \ge t$ such that $B + c \subseteq A$.*

*Proof* Let $I$ be an infinite interval such that $\mu_I(^*A) = \mathrm{BD}(A)$. It follows that $\mu_I(^*A + t), \mu_I(^*A + 2t), \ldots$ cannot all be pairwise almost everywhere disjoint, whence there are $k \le l$ such that $\mu_I((^*A + kt) \cap (^*A + lt)) > 0$, whence

---

[2]Indeed, if $A$ is piecewise syndetic, then $A + [0, k]$ is thick for some $k \in \mathbb{N}$. Thick sets are easily seen to contain FS-sets, whence, by the Strong version of Hindman's theorem (Corollary 8.7), $A+i$ contains an FS-set for some $i \in [0, k]$. It follows immediately that $A$ has the sumset property.

$BD((A + kt) \cap (A + lt)) > 0$. Let $c := |k - l|t$, so $BD(A \cap (A + c)) > 0$. Let $B := (A \cap (A + c)) - c$. Then this $B$ and $c$ are as desired.

After Nathanson's result, there had been very little progress made on proving Conjecture 13.1. In 2015, Di Nasso, Goldbring, Leth, Lupini, Jin, and Mahlburg proved the following result [42]:

**Theorem 13.6**

1. If $BD(A) > \frac{1}{2}$, then $A$ has the sumset property.
2. If $BD(A) > 0$, then there is $k \in \mathbb{N}$ such that $A \cup (A + k)$ has the sumset property.

In the same paper, the authors establish that *pseudorandom* sets also satisfy the sumset property. Very recently, significantly building upon the ideas from [42], Moreira, Richter, and Robertson proved Conjecture 13.1 in a very strong form [102]:

**Theorem 13.7** *If $BD(A) > 0$, then $A$ has the sumset property.*

The proof of the previous theorem is significantly beyond the scope of this book. However, we believe that there is value in giving the proof of Theorem 13.6 as it is a perfect example of the utility of nonstandard techniques in combinatorial number theory; the proof will be given in the next section.

We end this section by establishing a nonstandard reformulation of the sumset property. We will actually need the following more general statement:

**Proposition 13.8** *Given $A \subseteq \mathbb{N}$ and $k \in \mathbb{Z}$, the following are equivalent:*

1. *there exists $B = \{b_1 < b_2 < \cdots\}$ and $C = \{c_1 < c_2 < \cdots\}$ such that $b_i + c_j \in A$ for $i \leq j$ and $b_i + c_j \in A + k$ for $i > j$;*
2. *there exist nonprincipal ultrafilters $\mathcal{U}$ and $\mathcal{V}$ on $\mathbb{N}$ such that $A \in \mathcal{U} \oplus \mathcal{V}$ and $A + k \in \mathcal{V} \oplus \mathcal{U}$;*
3. *there exist infinite $\beta, \gamma \in {}^*\mathbb{N}$ such that $\beta + {}^*\gamma \in {}^{**}A$ and $\gamma + {}^* \beta \in {}^{**}A + k$.*

*Proof* First suppose that (1) holds as witnessed by $B$ and $C$. By assumption, the collection of sets

$$\{B\} \cup \{A - c : c \in C\}$$

has the finite intersection property with the Frechét filter, whence there is a non-principal ultrafilter $\mathcal{U}$ on $\mathbb{N}$ extending this family. Likewise, there is a nonprincipal ultrafilter $\mathcal{V}$ on $\mathbb{N}$ extending the family $\{C - k\} \cup \{A - b : b \in B\}$. These $\mathcal{U}$ and $\mathcal{V}$ are as desired.

Next, given (2), take $\beta, \gamma \in {}^*\mathbb{N}$ such that $\mathcal{U} = \mathcal{U}_\beta$ and $\mathcal{V} = \mathcal{U}_\gamma$. These $\beta$ and $\gamma$ are as desired.

Finally, suppose that $\beta, \gamma \in {}^*\mathbb{N}$ are as in (3). We define $B = \{b_1 < b_2 < b_3 < \cdots\}$ and $C = \{c_1 < c_2 < c_3 < \cdots\}$ recursively as follows. Suppose that $b_i$ and $c_j$ for $i, j = 1, \ldots, n$ have been constructed so that, for all $i, j$ we have:

- $b_i + c_j \in A$ if $i \leq j$;
- $b_i + c_j \in A + k$ if $i > j$;

- $b_i + \gamma \in {}^*A$;
- $c_j + \beta \in {}^*A + k$.

Applying transfer to the statement "there is $x \in {}^*\mathbb{N}$ such that $x + c_j \in {}^*A + k$ for $j = 1, \ldots, n$ and $x > b_n$ and $x + {}^*\gamma \in {}^{**}A$" (which is witnessed by $\beta$), we get $b_{n+1} \in \mathbb{N}$ such that $b_{n+1} > b_n$, $b_{n+1} + c_j \in A + k$ for $j = 1, \ldots, n$ and for which $b_{n+1} + \gamma \in {}^*A$. Next, apply transfer to the statement "there is $y \in {}^*\mathbb{N}$ such that $b_i + y \in {}^*A$ for $i = 1, \ldots, n + 1$ and $y > c_n$ and $y + {}^*\beta \in {}^{**}A + k$" (which is witnessed by $\gamma$), we get $c_{n+1} \in \mathbb{N}$ such that $c_{n+1} > c_n$ and for which $b_i + c_{n+1} \in A$ for $i = 1, \ldots, n + 1$ and for which $c_{n+1} + \beta \in {}^*A$. This completes the recursive construction.

Taking $k = 0$ in the previous proposition yields a nonstandard reformulation of the sumset property.

**Corollary 13.9** *Given $A \subseteq \mathbb{N}$, the following are equivalent:*

1. *$A$ has the sumset property;*
2. *there exist nonprincipal ultrafilters $\mathcal{U}$ and $\mathcal{V}$ on $\mathbb{N}$ such that $A \in (\mathcal{U} \oplus \mathcal{V}) \cap (\mathcal{V} \oplus \mathcal{U})$;*
3. *there exist infinite $\xi, \eta \in {}^*\mathbb{N}$ such that $\xi + {}^*\eta, \eta + {}^*\xi \in {}^{**}A$.*

## 13.2   A 1-Shift Version of Erdős' Conjecture

In this section, we prove Theorem 13.6. We first show how the first part of that theorem, together with Proposition 13.8, yields the second item of the theorem, which we state in an even more precise form.

**Proposition 13.10** *Suppose that $\mathrm{BD}(A) > 0$. Then there exists $B = \{b_1 < b_2 < \cdots\}$, $C = \{c_1 < c_2 < \cdots\}$, and $k \in \mathbb{N}$ such that $b_i + c_j \in A$ for $i \leq j$ and $b_i + c_j \in A + k$ for $i > j$.*

*Proof* By Proposition 10.10, we may fix $n \in \mathbb{N}$ such that $\mathrm{BD}(A + [-n, n]) > \frac{1}{2}$. By Theorem 13.6(1) and Corollary 13.9, we may take infinite $\beta, \gamma \in {}^*\mathbb{N}$ such that $\beta + {}^*\gamma, \gamma + {}^*\beta \in {}^{**}A + [-n, n]$. Take $i, j \in [-n, n]$ such that $\beta + {}^*\gamma \in {}^{**}A + i$ and $\gamma + {}^*\beta \in {}^{**}A + j$. Without loss of generality, $i < j$. Set $k := j - i$. Then $\beta + {}^*(\gamma - i) \in {}^{**}A$ and $(\gamma - i) + {}^*\beta \in {}^{**}A + k$, whence the conclusion holds by Proposition 13.8.

In order to prove the first item in Theorem 13.6, we need one technical lemma:

**Lemma 13.11** *Suppose that $\mathrm{BD}(A) = r > 0$. Suppose further that $(I_n)$ is a sequence of intervals with witnessing the Banach density of $A$. Then there is $L \subseteq \mathbb{N}$ satisfying:*

1. *$\limsup_{n \to \infty} \frac{|L \cap I_n|}{|I_n|} \geq r$;*
2. *for all finite $F \subseteq L$, $A \cap \bigcap_{x \in F}(A - x)$ is infinite*

*Proof* First, we note that it suffices to find $L$ satisfying (1) and

(2') there is $x_0 \in {}^*A \setminus A$ such that $x_0 + L \subseteq {}^*A$.

Indeed, given finite $F \subseteq L$ and $K \subseteq \mathbb{N}$, $x_0$ witnesses the truth of "there exists $x \in {}^*A$ such that $x + F \subseteq {}^*A$ and $x \notin K$" whence, by transfer, such an $x$ can be found in $\mathbb{N}$, establishing (2).

In the rest of the proof, we fix infinite $H \in {}^*\mathbb{N}$ and let $\mu$ denote Loeb measure on $I_H$. In addition, for any $\alpha \in {}^*\mathbb{N}$ and hyperfinite $X \subseteq {}^*\mathbb{N}$, we set $d_\alpha(X) := \frac{|X|}{|I_\alpha|}$. Finally, we fix $\epsilon \in (0, \frac{1}{2})$.

Next we remark that it suffices to find a sequence $X_1, X_2, \ldots$ of internal subsets of $I_H$ and an increasing sequence $n_1 < n_2 < \cdots$ of natural numbers such that, for each $i$, we have:

(i)  $\mu(X_i) \geq 1 - \epsilon^i$ and,
(ii)  for each $x \in X_i$, we have $d_{n_i}({}^*A \cap (x + I_{n_i})) \geq r - \frac{1}{i}$.

Indeed, suppose that this has been accomplished and set $X := \bigcap_i X_i$. Then $X$ is Loeb measurable and $\mu(X) > 0$. Fix $y_0 \in X \setminus \mathbb{N}$ arbitrary and set $x_0$ to be the minimum element of ${}^*A$ that is greater than or equal to $y_0$; note that $x_0 - y_0 \in \mathbb{N}$ since $y_0 \in X$. Set $L := ({}^*A - x_0) \cap \mathbb{N}$; note that (2') is trivially satisfied. To see that (1) holds, note that

$$\limsup_{i \to \infty} d_{n_i}(L \cap I_{n_i}) = \limsup_{i \to \infty} d_{n_i}({}^*A \cap (x_0 + I_{n_i})) = \limsup_{i \to \infty} d_{n_i}({}^*A \cap (y_0 + I_{n_i})) \geq r,$$

where the last inequality follows from the fact that $y_0 \in X$.

Thus, to finish the lemma, it suffices to construct the sequences $(X_i)$ and $(n_i)$. Suppose that $X_1, \ldots, X_{i-1}$ and $n_1 < \cdots < n_{i-1}$ have been constructed satisfying the conditions above. For $\alpha \in {}^*\mathbb{N}$, set

$$Y_\alpha := \{x \in I_H \;:\; d_\alpha({}^*A \cap (x + I_m)) \geq r - \frac{1}{i}\}.$$

Set $Z := \{\alpha \in {}^*\mathbb{N} \;:\; n_{i-1} < \alpha \text{ and } d_H(Y_\alpha) > 1 - \epsilon^i\}$. Note that $Z$ is internal. It will be enough to show that $Z$ contains all sufficiently small infinite elements of ${}^*\mathbb{N}$, for then, by underflow, there is $n_i \in Z \cap \mathbb{N}$. Setting $X_i := Y_{n_i}$, these choices of $X_i$ and $n_i$ will be as desired.

We now work towards proving that $Z$ contains all sufficiently small infinite elements of ${}^*\mathbb{N}$. First, we remark that we may assume, without loss of generality, that the sequences $(|I_n|)$ and $(b_n)$ are increasing, where $b_n$ denotes the right endpoint of $I_n$. Fix $K \in {}^*\mathbb{N} \setminus \mathbb{N}$ such that $2b_K/|I_H| \approx 0$. We finish the proof of the lemma by proving that $K \in Z$, which we claim follows from the following two facts:

(a)  for all $x \in I_H$, $\mathrm{st}(d_K({}^*A \cap (x + I_K))) \leq r$;
(b)  $\frac{1}{|I_H|} \sum_{x \in I_H} d_K({}^*A \cap (x + I_K)) \approx r$.

To see that these facts imply that $K \in Z$, for $x \in I_H$, set $f(x) := d_K(^*A \cap (x + I_K))$. It is enough to show that $f(x) \approx r$ for $\mu$-almost all $x \in I_H$. Given $n$, let $A_n := \{x \in I_H : f(x) < r - \frac{1}{n}\}$. Suppose, towards a contradiction, that $\mu(A_n) = s > 0$. By (a), we may fix a positive infinitesimal $\eta$ such that $f(x) \leq r + \eta$ for all $x \in I_H$. We then have

$$\frac{1}{|I_H|} \sum_{x \in I_H} f(x) = \frac{1}{|I_H|} \left[ \sum_{x \in A_n} f(x) + \sum_{x \notin A_n} f(x) \right] < s(r - \frac{1}{n}) + (1 - s)(r + \eta).$$

Since the right-hand side of the above display is appreciably less than $s$, we get a contradiction to (b).

It remains to establish (a) and (b). (a) follows immediately from the fact that $BD(A) = r$. To see (b), we first observe that

$$\frac{1}{|I_H|} \sum_{x \in I_H} d_K(^*A \cap (x + I_K)) = \frac{1}{|I_K|} \sum_{y \in I_K} \frac{1}{|I_H|} \sum_{x \in I_H} \chi_{^*A}(x + y).$$

Fix $y \in I_K$. Since $|\sum_{x \in I_H} \chi_{^*A}(x + y) - |^*A \cap I_H|| \leq 2y \leq 2b_K$, we have that

$$\left| \frac{1}{|I_H|} \sum_{x \in I_H} \chi_{^*A}(x + y) - d_H(^*A) \right| \approx 0.$$

Since a hyperfinite average of infinitesimals is infinitesimal, we see that

$$\frac{1}{|I_H|} \sum_{x \in I_H} d_K(^*A \cap (x + I_K)) \approx \frac{1}{|I_K|} \sum_{y \in I_K} d_H(^*A) \approx r,$$

establishing (b). $\square$

*Remark 13.12* A significant strengthening of the previous lemma was one of the main ingredients in the full resolution of Conjecture 13.1 given in [102].

*Proof of Theorem 13.6* Set $r := BD(A)$. Let $(I_n)$ witness the Banach density of $A$ and let $L := (l_n)$ be as in the previous lemma. We may then define an increasing sequence $D := (d_n)$ contained in $A$ such that $l_i + d_n \in A$ for $i \leq n$.[3] Now take $N$ such that $\mu_{I_N}(^*L) \geq r$. Note also that $\mu_{I_N}(^*A - d_n) \geq r$ for any $n$. Since $r > 1/2$, for any $n$ we have that $\mu_{I_N}(^*L \cap (^*A - d_n)) \geq 2r - 1 > 0$. By a standard measure theory fact, by passing to a subsequence of $D$ if necessary, we may assume that, for

---

[3]Notice that at this point we have another proof of Nathanson's Theorem 13.4: if we set $B := \{d_n, d_{n+1}, \ldots\}$ and $C := \{l_1, \ldots, l_n\}$, then $B + C \subseteq A$.

each $n$, we have that $\mu_{I_N}(^*L \cap \bigcap_{i \leq n}(^*A - d_i)) > 0$. In particular, for every $n$, we have that $L \cap \bigcap_{i \leq n}(A - d_i)$ is infinite.

We may now conclude as follow. Fix $b_1 \in L$ arbitrary and take $c_1 \in D$ such that $b_1 + c_1 \in A$. Now assume that $b_1 < \cdots < b_n$ and $c_1 < \cdots < c_n$ are taken from $L$ and $D$ respectively such that $b_i + c_j \in A$ for all $i, j = 1, \ldots, n$. By assumption, we may find $b_{n+1} \in L \cap \bigcap_{i \leq n}(A - c_i)$ with $b_{n+1} > b_n$ and then we may take $c_{n+1} \in D$ such that $b_i + c_{n+1} \in A$ for $i = 1, \ldots, n+1$.

## 13.3 A Weak Density Version of Folkman's Theorem

At the beginning of this chapter, we discussed the fact that the density version of Hindman's theorem is false. In fact, the odd numbers also show that the density version of Folkman's theorem is also false. (Recall that Folkman's theorem stated that for any finite coloring of $\mathbb{N}$, there are arbitrarily large finite sets $G$ such that $FS(G)$ are monochromatic.) However, we can use Lemma 13.11 to prove a weak density version of Folkman's theorem. Indeed, the proof of Lemma 13.11 yields the following:

**Lemma 13.13** *Suppose that $A \subseteq \mathbb{N}$ is such that $BD(A) \geq r$. Then there is $\alpha \in {}^*A \setminus A$ such that $BD(A - \alpha) \geq r$.*

One should compare the previous lemma with Beiglbock's Lemma 12.15. Indeed, a special case of (the nonstandard formulation of) Lemma 12.15 yields $\alpha \in {}^*\mathbb{N} \setminus \mathbb{N}$ such that $BD(^*A - \alpha) \geq BD(A)$; the previous lemma is stronger in that it allows us to find $\alpha \in {}^*A$. We can now prove the aforementioned weak version of a density Folkman theorem.

**Theorem 13.14** *Fix $k \in \mathbb{N}$ and suppose $A \subseteq \mathbb{N}$ is such that $BD(A) > 0$. Then there exist increasing sequences $(x_n^{(i)})$ for $i = 0, 1, 2, \ldots, k$ such that, for any $i$ and any $n_i \leq n_{i+1} \leq \cdots \leq n_k$, we have $x_{n_i}^{(i)} + x_{n_{i+1}}^{(i+1)} + \cdots + x_{n_k}^{(k)} \in A$.*

The reason we think of the previous theorem as a weak density version of Folkman's theorem is that if all of the sequences were identical, then we would in particular have a set of size $k$ all of whose finite sums belong to $A$.

*Proof of Theorem 13.14* Set $A = A^{(k)}$. Repeatedly applying Lemma 13.13, one can define, for $i = 0, 1, \ldots, k$, subsets $A^{(i)}$ of $\mathbb{N}$ and $\alpha_i \in {}^*A^{(i)}$ such that $A^{(i)} + \alpha_{i+1} \subseteq {}^*A^{(i+1)}$ for all $i < k$. We then define the sequences $(x_n^{(i)})$ for $i = 0, 1, 2, \ldots, k$ and finite subsets $A_n^{(i)}$ of $A^{(i)}$ so that:

- for $i = 0, 1, \ldots, k$ and any $n$, we have $x_n^{(i)} \in A_n^{(i)}$,
- for $i = 0, 1, \ldots, k$ and any $n \leq m$, we have $A_n^{(i)} \subseteq A_m^{(i)}$, and
- for $i = 0, 1, \ldots, k-1$ and any $n \leq m$, we have $A_n^{(i)} + x_m^{(i+1)} \subseteq A_m^{(i+1)}$.

It is clear that the sequences $(x_n^{(i)})$ defined in this manner satisfy the conclusion of the theorem. Suppose that the sequences $(x_n^{(i)})$ and $A_n^{(i)}$ have been defined for $n < m$. We now define $x_m^{(i)}$ and $A_m^{(i)}$ by recursion for $i = 0, 1, \ldots, k$. We set $x_m^{(0)}$ to be any member of $A^{(0)}$ larger than $x_{m-1}^{(0)}$ and set $A_m^{(0)} := A_{m-1}^{(0)} \cup \{x_{m-1}^{(0)}\}$. Supposing that the construction has been carried out up through $i < k$, by transfer of the fact that $A_m^{(i)} + \alpha^{(i+1)} \subseteq {}^*A^{(i+1)}$, we can find $x_m^{(i+1)} \in A^{(i+1)}$ larger than $x_{m-1}^{(i+1)}$ such that $A_m^{(i)} + x_m^{(i+1)} \subseteq A^{(i+1)}$. We then define $A_m^{(i+1)} := A_{m-1}^{(i+1)} \cup (A_m^{(i)} + x_m^{(i+1)})$. This completes the recursive construction and the proof of the theorem.

The usual compactness argument gives a finitary version:

**Corollary 13.15** *Suppose that $k \in \mathbb{N}$ and $\varepsilon > 0$ are given. Then there exists $m$ such that for any interval $I$ of length at least $m$ and any subset $A$ of $I$ such that $|A| > \varepsilon |I|$, there exist $(x_n^{(i)})$ for $i, n \in \{0, 1, \ldots, k\}$ such that $x_{n_i}^{(i)} + x_{n_{i+1}}^{(i+1)} + \cdots + x_{n_{\ell-1}}^{(k)} \in A$ for any $i = 0, 1, \ldots, k$ and any $0 \le n_i \le n_{i+1} \le \cdots \le n_{\ell-1} \le k$.*

## Notes and References

The proof of Corollary 13.10 from Theorem 13.6 given in [42] proceeds via Ramsey's theorem. The ultrafilter reformulation of the sumset property was first observed by Di Nasso and was used to give this alternate derivation of Corollary 13.10 from Theorem 13.6. The paper [42] also presents a version of Theorem 13.6 and Corollary 13.10 for countable amenable groups. Likewise, the paper [102] presents a version of the solution to Conjecture 13.1 for countable amenable groups.

# Chapter 14
# Near Arithmetic Progressions in Sparse Sets

Szemerédi's theorem says that relatively dense sets contain arithmetic progressions. The purpose of this chapter is to present a result of Leth from [89] which shows that certain sparse sets contain "near" arithmetic progressions. We then detail the connection between the aforementioned theorem of Leth and the Erdős-Turán conjecture.

## 14.1 The Main Theorem

We begin by making precise the intuitive notion of "near" arithmetic progression mentioned in the introduction.

**Definition 14.1** Fix $w \in \mathbb{N}_0$ and $t, d \in \mathbb{N}$.[1] A $(t, d, w)$-*progression* is a set of the form

$$\mathrm{B}(b, t, d, w) := \bigcup_{i=0}^{t-1}[b + id, b + id + w].$$

By a *block progression* we mean a $(t, d, w)$-progression for some $t, d, w$.

Note that a $(t, d, 0)$-progression is the same thing as a $t$-term arithmetic progression with difference $d$.

---

[1] In this chapter, we deviate somewhat from our conventions so as to match up with the notation from [89].

© Springer Nature Switzerland AG 2019
M. Di Nasso et al., *Nonstandard Methods in Ramsey Theory and Combinatorial Number Theory*, Lecture Notes in Mathematics 2239, https://doi.org/10.1007/978-3-030-17956-4_14

**Definition 14.2** If $A \subseteq \mathbb{N}$, we say that $A$ *nearly contains a* $(t, d, w)$-*progression* if there is a $(t, d, w)$-progression $B(b, t, d, w)$ such that $A \cap [b + id, b + id + w] \neq \emptyset$ for each $i = 1, \ldots, t - 1$.

Thus, if $A$ nearly contains a $(t, d, 0)$-progression, then $A$ actually contains a $t$-term arithmetic progression. Consequently, when $A$ nearly contains a $(t, d, w)$-progression with "small" $w$, then this says that $A$ is "close" to containing an arithmetic progression. The main result of this chapter allows us to conclude that even relatively sparse sets with a certain amount of density regularity nearly contain block progressions satisfying a further homogeneity assumption that we now describe.

**Definition 14.3** Suppose that $A \subseteq \mathbb{N}$, $I$ is an interval in $\mathbb{N}$, and $0 < s < 1$. We say that $A$ *nearly contains a* $(t, d, w)$-*progression in $I$ with homogeneity $s$* if there is some $B(b, t, d, w)$ contained in $I$ such that the following two conditions hold for all $i, j = 0, 1, \ldots, t - 1$:

(i) $\delta(A, [b + id, b + id + w]) \geq (1 - s)\delta(A, I)$
(ii) $\delta(A, [b + id, b + id + w]) \geq (1 - s)\delta(A, [b + jd, b + jd + w])$.

Thus, for small $s$, we see that $A$ meets each block in a density that is roughly the same throughout and that is roughly the same as on the entire interval.

The density regularity condition roughly requires that on sufficiently large subintervals of $I$, the density does not increase too rapidly. Here is the precise formulation:

**Definition 14.4** Suppose that $I \subseteq \mathbb{N}$ is an interval, $r \in \mathbb{R}$, $r > 1$, and $m \in \mathbb{N}$. We say that $A \subseteq I$ has the $(m, r)$-*density property on $I$* if, whenever $J \subseteq I$ is an interval with $|J|/|I| \geq 1/m$, then $\delta(A, J) \leq r\delta(A, I)$.

Of course, given any $m \in \mathbb{N}$ and $A \subseteq I$, there is $r \in \mathbb{R}$ such that $A$ has the $(m, r)$-density property on $I$. The notion becomes interesting when we think of $r$ as fixed.

Given a hyperfinite interval $I \subseteq {}^*\mathbb{N}$, $r \in {}^*\mathbb{R}$, $r > 1$, and $M \in {}^*\mathbb{N}$, we say that an internal set $A \subseteq I$ has the *internal* $(M, r)$-*density property on $I$* if the conclusion of the definition above holds for internal subintervals $J$ of $I$.

**Lemma 14.5** *Suppose that $A \subseteq [1, N]$ is an internal set with the internal $(M, r)$-density property for some $M > \mathbb{N}$. Let $f : [0, 1] \to [0, 1]$ be the (standard) function given by*

$$f(x) := \mathrm{st}\left(\frac{|A \cap [1, xN]|}{|A \cap [1, N]|}\right).$$

*Then $f$ is a Lipschitz function with Lipschitz constant $r$.*

*Proof* Fix $x < y$ in $[0, 1]$. Write $x := \mathrm{st}(K/N)$ and $y := \mathrm{st}(L/N)$. Since $y - x \neq 0$, we have that $\frac{L-K}{N}$ is not infinitesimal; in particular, $\frac{L-K}{N} > 1/M$. Since $A$ has the $(M, r)$-density property on $[1, N]$, we have that $\delta(A, [K, L]) \leq r\delta(A, [1, N])$.

Thus, it follows that

$$f(y) - f(x) = \text{st}\left(\frac{|A \cap [K, L]|}{|A \cap [1, N]|}\right) = \text{st}\left(\delta(A, [K, L])\frac{L - K}{|A \cap [1, N]|}\right)$$

$$\leq r \, \text{st}\left(\frac{L - K}{N}\right) = r(y - x).$$

Here is the main result of this section:

**Theorem 14.6 (Leth)** *Fix functions* $g, h : \mathbb{R}^{>0} \to \mathbb{R}^{>0}$ *such that h is increasing and* $g(x) \to \infty$ *as* $x \to \infty$. *Fix also* $s > 0, r > 1,$ *and* $j, t \in \mathbb{N}$. *Then there is* $m = m(g, h, s, r, t, j) \in \mathbb{N}$ *such that, for all* $n > m$, *whenever I is an interval of length n and* $A \subseteq I$ *is nonempty and has the* $(m, r)$-*density property on I, then A contains a* $(t, d, w)$-*almost progression with homogeneity s such that* $w/d < h(d/n)$ *and* $1/g(m) < d/n < 1/j$.

Roughly speaking, if $A$ has sufficient density regularity, then $A$ contains an almost-progression with "small" $w$ (small compared to the distance of the progression).

The proof of the theorem relies on the following standard lemma; see [89, Lemma 1].

**Lemma 14.7** *Suppose that* $E \subseteq \mathbb{R}$ *has positive Lebesgue measure and* $t \in \mathbb{N}$. *Then there is* $v > 0$ *such that, for all* $0 < u < v$, *there is an arithmetic progression in E of length t and difference u.*

We stress that in the previous lemma, $u$ and $v$ are real numbers.

*Proof of Theorem 14.6* Fix $g, h, s, r, j, t$ as in the statement of Theorem 14.6. We show that the conclusion holds for all infinite $M$, whence by underflow there exists $m \in \mathbb{N}$ as desired. Thus, we fix $M > \mathbb{N}$ and consider $N > M$, an interval $I \subseteq {}^*\mathbb{N}$ of length $N$, and a hyperfinite subset $A \subseteq I$ that has the internal $(M, r)$-density property on $I$. Without loss of generality, we may assume that $I = [1, N]$. Suppose that we can find $B, D, W \in {}^*\mathbb{N}$ and standard $c > 0$ such that $[B, B + (t - 1)D + W] \subseteq [1, N]$ and, for all $i = 0, 1, \ldots, t - 1$, we have:

$$\delta(A, [1, N])(c - \frac{s}{2}) \leq \delta(A, [B + iD, B + iD + W]) \leq \delta(A, [1, N])(c + \frac{s}{4}). \quad (\dagger)$$

We claim that $A$ nearly contains the internal $(t, D, W)$-progression $B(B, t, D, W)$ with homogeneity $s$. Indeed, item (i) of Definition 14.3 is clear. For item (ii), observe that

$$\delta(A, [B + iD, B + iD + W]) \geq \delta(A, [1, N])(c - \frac{s}{2})$$

$$\geq \delta(A, [B + jD, B + jD + W])(\frac{c - \frac{s}{2}}{c + \frac{s}{4}})$$

and note that $\frac{c - \frac{s}{2}}{c + \frac{s}{4}} > 1 - s$. Thus, it suffices to find $B, D, W, c$ satisfying (†) and for which $W/D < h(D/N)$ and $1/g(M) < D/N < 1/j$.

Let $f$ be defined as in the statement of Lemma 14.5. Set $b := \mathrm{st}(B/N)$, $d := \mathrm{st}(D/N)$, and $w := \mathrm{st}(W/N)$. Assume that $w \neq 0$. Then we have that

$$\mathrm{st}\left(\frac{\delta(A, [B + iD, B + iD + W])}{\delta(A, [1, N])}\right) = \frac{f(b + id + w) - f(b + id)}{w}.$$

We thus want to find $B, D, W$ and $c$ satisfying

$$c - \frac{s}{2} < \frac{f(b + id + w) - f(b + id)}{w} < c + \frac{s}{4}. \qquad (\dagger\dagger)$$

Now the middle term in (††) looks like a difference quotient and the idea is to show that one can bound $f'(b + id)$ for $i = 0, 1, \ldots, t - 1$. Indeed, by Lemma 14.5, $f$ is Lipschitz, whence it is absolutely continuous. In particular, by the Fundamental Theorem of Calculus, $f$ is differentiable almost everywhere and $f(x) = \int_0^x f'(u)du$. Since $f(0) = 0$ and $f(1) = 1$, it follows that $\{x \in [0, 1] : f'(x) \geq (1 - \frac{s}{4})\}$ has positive measure. In particular, there is $c > 1$ such that

$$E := \{x \in [0, 1] : c - \frac{s}{4} \leq f'(x) \leq c\}$$

has positive measure. By Lemma 14.7, there is $b \in E$ and $0 < u < 1/j$ such that $b, b + u, b + 2u, \ldots, b + (t - 1)u \in E$. Take $B, D \in [1, N]$ such that $b = \mathrm{st}(B/N)$ and $u = \mathrm{st}(D/N)$. Note that $g(M)$ is infinite and $D/N$ is noninfinitesimal, so $1/g(M) < D/N < 1/j$. It remains to choose $W$. Since $f$ is differentiable on $E$, there is $w > 0$ sufficiently small so that for all $i = 0, 1, \ldots, t - 1$, we have $|f'(b + id) - \frac{f(b+id+w)-f(b+id)}{w}| < \frac{s}{4}$. For this $w$, (††) clearly holds; we now take $W$ such that $w = \mathrm{st}(W/N)$. Since $h(D/N)$ is nonfinitesimal (as $D/N$ is noninfinitesimal), if $w$ is chosen sufficiently small, then $W/D < h(D/N)$.

Theorem 14.6 implies a very weak form of Szemeredi's theorem.

**Corollary 14.8** *Suppose that* $\mathrm{BD}(A) > 0$. *Suppose that* $g, h, s, t, j$ *are as in the hypothesis of Theorem 14.6. Then for $n$ sufficiently large, there is an interval $I$ of length $n$ such that $A \cap I$ contains a $(t, s, d)$-almost progression in $I$ with $w/d < h(d/n)$ and $1/g(m) < d/n < 1/j$.*

*Proof* Take $r \in \mathbb{R}$ with $r > 1$ satisfying $\mathrm{BD}(A) > 1/r$. Let $m := m(g, h, s, r, t, j)$ as in the conclusion of Theorem 14.6. Let $n > m$ and take an interval $I$ of length $n$ such that $\delta(A, I) > 1/r$. It remains to observe that $A \cap I$ has the $(m, r)$-density property on $I$.

## 14.2 Connection to the Erdős-Turán Conjecture

Leth's original motivation was the following conjecture of Erdős and Turán from [47]:

*Conjecture 14.9 (Erdős-Turán)* Suppose that $A = (a_n)$ is a subset of $\mathbb{N}$ such that $\sum 1/a_n$ diverges. Then $A$ contains arbitrarily long arithmetic progressions.

Leth first observed the following standard fact about the densities of sequences satisfying the hypotheses of the Erdős-Turán conjecture.

**Lemma 14.10** *Suppose that $A = (a_n)$ is enumerated in increasing order and is such that $\sum 1/a_n$ diverges. Then, for arbitrarily large $n$, one has $\delta(A, n) > 1/(\log n)^2$.*

*Proof* We argue by contrapositive. Suppose that $\delta(A, n) \leq 1/(\log n)^2$ for all $n \geq n_0 \geq 4$. We first show that this implies that $a_n \geq \frac{1}{2}n(\log n)^2$ for all $n > n_0$. Suppose otherwise and fix $n \geq n_0$. Then $|A \cap [1, \frac{1}{2}n(\log n)^2]| \geq n$. On the other hand, by our standing assumption, we have that

$$|A \cap [1, \frac{1}{2}n(\log n)^2]) \leq \frac{1/2n(\log n)^2}{(\log((1/2n(\log n))^2)} \leq \frac{1}{2}n,$$

yielding the desired contradiction.

Since $a_n \geq \frac{1}{2}n(\log n)^2$ eventually, we have that

$$\sum \frac{1}{a_n} \leq \sum \frac{2}{n(\log n)^2},$$

whence $\sum \frac{1}{a_n}$, converges.

The truth of the following conjecture, together with the theorem that follows it, would imply that, for sets satisfying the density condition in the previous lemma, the existence of almost arithmetic progressions implies the existence of arithmetic progressions.

*Conjecture 14.11 (Leth)* Fix $t \in \mathbb{N}$ and $c > 0$. Then there is $n_0 := n_0(t, c)$ such that, for all $n \geq n_0$, whenever $A \subseteq \mathbb{N}$ is such that $\delta(A, n) > 1/(c \log n)^{2 \log \log n}$, then $A$ nearly contains a $(t, d, w)$-progression on $[1, n]$ with $w/d < d/n$ where $d$ is a power of 2.

We should remark that requiring that $d$ be a power of 2 is not much of an extra requirement. Indeed, our proof of Theorem 14.6 shows that one can take $d$ there to be a power of 2. For any $t$ and $c$, we let $L(t, c)$ be the statement that the conclusion of the previous conjecture holds for the given $t$ and $c$. We let $L(t)$ be the statement that $L(t, c)$ holds for all $c > 0$.

**Theorem 14.12** *Suppose that $L(t)$ is true for a given $t \in \mathbb{N}$. Further suppose that $A \subseteq \mathbb{N}$ is such that there is $c > 0$ for which, for arbitrarily large $n$, one has $\delta(A, n) > c/(\log n)^2$. Then $A$ contains an arithmetic progression of length $t$.*

Before we prove this theorem, we state the following standard combinatorial fact, whose proof we leave as an exercise to the reader (alternatively, this is proven in [89, Proposition 1]).

**Proposition 14.13** *Let $m, n \in \mathbb{N}$ be such that $m < n$, let $A \subseteq \mathbb{N}$, and let $I$ be an interval of length $n$. Then there is an interval $J \subseteq I$ of length $m$ such that $\delta(A, J) > \delta(A, I)/2$.*

*Proof of Theorem 14.12* For reasons that will become apparent later in the proof, we will need to work with the set $2A$ rather than $A$. Note that $2A$ satisfies the hypothesis of the theorem for a different constant $c' > 0$.

By overflow, we may find $M > \mathbb{N}$ such that $\delta(^*(2A), M) > \frac{c'}{(\log M)^2}$. Take $L > \mathbb{N}$ such that $2^{2^L} \leq M < 2^{2^{L+1}}$ and set $N := 2^{2^L}$. If we apply Proposition 14.13 to any $n \leq N$ and $I = [1, N]$, we can find an interval $I_n \subseteq [1, M]$ of length $n$ such that

$$|^*(2A) \cap I_n| > \frac{c'M}{2(\log M)^2} \geq \frac{c'M}{2(\log 2^{2^{L+1}})^2} = \frac{c'/8}{(\log N)^2}.$$

For $1 \leq k \leq L$, write $I_{2^{2^k}} = [x_k, y_k]$.

We will now construct an internal set $B \subseteq [1, N]$ such that $\delta(B, N) > \frac{1}{(c'' \log N)^{2 \log \log N}}$, where $c'' := \sqrt{8/c'}$. Since we are assuming that $L(t)$ holds, by transfer we will be able to find an internal $(t, d, w)$-progression nearly inside of $B$ with $w/d < d/N$ and $w$ and $d$ both powers of 2. The construction of $B$ will allow us to conclude that $^*(2A)$ contains a $t$-termed arithmetic progression of difference $d$, whence so does $2A$ by transfer, and thus so does $A$.

Set $B_0 := [1, N]$ and, for the sake of describing the following recursive construction, view $B_0$ as the union of two subintervals of length $N/2 = 2^{2^L - 1} = 2^{2^L - 2^0}$; we refer to these subintervals of $B_0$ as *blocks*. Now divide each block in $B_0$ into $2 = 2^{2^0}$ intervals of length $2^{2^L - 2^0}/2^{2^0} = 2^{2^L - 2^1}$ and, for each $0 \leq j < 2^{2^0}$, we place the $j^{\text{th}}$ subblock of each block in $B_0$ into $B_1$ if and only if $x_0 + j \in {}^*2A$.

Now divide each block in $B_1$ into $2^{2^1}$ intervals of length $2^{2^L - 2^1}/2^{2^1} = 2^{2^L - 2^2}$ and, for each $0 \leq j < 2^{2^1}$, we place the $j$th subblock of each block in $B_1$ into $B_2$ if and only if $x_1 + j \in {}^*2A$.

We continue recursively in this manner. Thus, having constructed the hyperfinite set $B_k$, which is a union of blocks of length $2^{2^L - 2^k}$, we break each block of $B_k$ into $2^{2^k}$ many intervals of length $2^{2^L - 2^k}/2^{2^k} = 2^{2^L - 2^{k+1}}$ and we place the $j$th subblock of each block in $B_k$ into $B_{k+1}$ if and only if $x_k + j \in {}^*2A$.

We set $B := B_L$. Since $|B_{k+1}|/|B_k| > \frac{c'/8}{(\log N)^2}$ for each $0 \le k < L$, it follows that

$$|B| > \frac{(c'/8)^L N}{(\log N)^{2L}} = \frac{N}{(c'' \log N)^{2 \log \log N}} .$$

By applying transfer to $L(t)$, we have that $B$ nearly contains an internal $(t, d, w)$-progression $B(b, t, d, w)$ contained in $[1, N]$ such that $w/d < d/N$ and $d$ is a power of 2. Take $k$ such that $2^{2^L - 2^{k+1}} \le d < 2^{2^L - 2^k}$. Note that this implies that $2^{2^L - 2^{k+1}} \mid d$. Also, we have

$$w < (d/N) \cdot d < (2^{-2^k}) 2^{2^L - 2^k} = 2^{2^L - 2^{k+1}} .$$

We now note that $B(b, t, d, w)$ must be contained in a single block $C$ of $B_k$. Indeed, since $d \mid 2^{2^L - 2^k}$ and $w \mid 2^{2^L - 2^{k+1}}$, we have $d + w < (\frac{1}{2} + \frac{1}{2^{2^k}})(2^{2^L - 2^k})$, whence the fact that $[b, b + w]$ and $[b + d, b + d + w]$ both intersect $B_k$ would imply that $[x_{k-1}, y_{k-1}]$ contains consecutive elements of $^*2A$, which is clearly a contradiction.

Now write $d = m \cdot 2^{2^L - 2^{k+1}}$. Take $0 \le j < 2^{2^k}$ so that $[b, b + w]$ intersects $B_{k+1}$ in the $j$th subblock of $C$ so $x_k + j \in {}^*2A$. Since $[b + d, b + d + w] \cap B_{k+1} \ne \emptyset$, we have that at least one of $x_k + j + (m - 1)$, $x_k + j + m$, or $x_k + j + (m + 1)$ belong to $^*(2A)$. However, since $x_k + j$ and $m$ are both even, it follows that we must have $x_k + j + m \in {}^*(2A)$. Continuing in this matter, we see that $x_k + j + im \in {}^*2A$ for all $i = 0, 1, \ldots, t - 1$. It follows by transfer that $2A$ contains a $t$-term arithmetic progression, whence so does $A$.

Putting everything together, we have:

**Corollary 14.14** *The Erdős-Turán conjecture follows from Leth's Conjecture.*

Leth used Theorem 14.6 to prove the following theorem, which is similar in spirit to Conjecture 14.6, except that it allows sparser sequences but in turn obtains almost progressions with weaker smallness properties relating $d$ and $w$.

**Theorem 14.15** *Suppose that $s > 0$ and $t \in \mathbb{N}^{>2}$ are given. Further suppose that $h$ is as in Theorem 14.6. Let $A \subseteq \mathbb{N}$ be such that, for all $\epsilon > 0$, we have $\delta(A, n) > 1/n^\epsilon$ for sufficiently large $n$. Then for sufficiently large $n$, $A$ nearly contains an $(t, d, w)$-progression on $[1, n]$ of homogeneity $s$ with $w/d < h(\log d / \log n)$, where $d$ is a power of 2.*

*Proof* Suppose that the conclusion is false. Then there is $N$ such that $^*A$ does not nearly contain any internal $(t, d, w)$-progression on $[1, N]$ of homogeneity $s$ with $w/d < h(\log d / \log N)$. It suffices to show that there is $\epsilon > 0$ such that $\delta(^*A, N) < 1/N^\epsilon$. Let $m$ be as in the conclusion of Theorem 14.6 with $r = 2$ and $g(x) = x$ (and $h$ as given in the assumptions of the current theorem).

*Claim* If $I \subseteq [1, N]$ is a hyperfinite interval with $|I| > \sqrt{N}$, then $^*A$ does *not* have the $(m, 2)$-density property on $I$.

We will return to the proof of the claim in a moment. We first see how the claim allows us to complete the proof of the theorem. Let $K > \mathbb{N}$ be the maximal $k \in {}^*\mathbb{N}$ such that $m^{2k} \leq N$, so $m^{2K} \leq N < m^{2K+2}$. We construct, by internal induction, for $i = 0, 1, \ldots, K$, a descending chain of hyperfinite subintervals $(I_i)$ of $I$ of length $m^{2K-i}$ as follows. By Proposition 14.13, we may take $I_0$ to be any hyperfinite subinterval of $I$ of length $m^{2K}$ such that $\delta(^*A, I_0) \geq \delta(^*A, N)/2$. Suppose that $i < K$ and $I_i$ has been constructed such that $|I_i| = m^{2k-i}$. Since $^*A$ does not have the $(m, 2)$ density property on $I_i$, there is a subinterval $I_{i+1}$ of length $|I_i|/m^{2k-i-1}$ with $\delta(^*A, I_{i+1}) \geq 2\delta(^*A, I_i)$. Notice now that $I_K$ is a hyperfinite interval of length $m^K \leq \sqrt{N} < m^{K+1}$ and $\delta(^*A, I_K) \geq 2^K \delta(^*A, I_0)$. It follows that

$$\delta(^*A, N) \leq 2\delta(A, I_0) \leq 2^{-(K-1)}\delta(A, I_K) \leq 2^{-(K-1)}.$$

It follows that

$$|A \cap [1, N]| \leq 2^{-(K-1)}N \leq 2^{-(K-1)}m^{2K+2} = m^{2K+2-(K-1)\frac{\log 2}{\log m}} = (m^{2K})^{1-z}.$$

if we set $z := \frac{(K-1)\log 2}{2K \log m} - \frac{1}{K}$. If we set $\epsilon := \mathrm{st}(z/2) = \frac{\log 2}{4 \log m}$, then it follows that $|A \cap [1, N]| \leq N^{1-\epsilon}$, whence this $\epsilon$ is as desired.

We now prove the claim. Suppose, towards a contradiction, that $I \subseteq [1, N]$ is a hyperfinite interval with $|I| > \sqrt{N}$ and is such that $^*A$ does have the $(m, 2)$-density property on $I$. By the choice of $m$, $^*A$ nearly contains an internal $(t, d, w)$-almost progression of homogeneity $s$ with $w/d < h(d/|I|)$ and $d > |I|/m > \sqrt{N}/m$. Notice now that $\mathrm{st}\left(\frac{\log d}{\log N}\right) \geq \mathrm{st}\left(\frac{1/2 \log N - \log m}{\log N}\right) = \frac{1}{2}$. Note that we trivially have that $d/|I| < 1/t$, whence $d/|I| < \log d/\log N$; since $h$ is increasing, we have that $w/d < h(\log d/\log N)$, contradicting the choice of $N$. This proves the claim and the theorem.

In [90, Theorem 3], Leth shows that one cannot replace $(\log d)/(\log n)$ with $d/n$ in the previous theorem.

## Notes and References

There are other generalizations of arithmetic progressions appearing in the literature, e.g. the notion of *quasi-progression* appearing in [131]. It should be noted that they use the term $(t, d, w)$-progression in a related, but different, manner than it is used in this chapter. The Erdős-Turan conjecture, first formulated in [48], is one of the most important open problems in combinatorial number theory. A positive solution would immediately generalize both Szemeredi's Theorem and the Green-Tao theorem on the existence of arbitrarily long arithmetic progressions in the primes [64].

# Chapter 15
# The Interval Measure Property

In this chapter, we define the notion, due to Leth, of internal subsets of the hypernatural numbers with the Interval Measure Property. Roughly speaking, such sets have a tight relationship between sizes of gaps of the set on intervals and the Lebesgue measure of the image of the set under a natural projection onto the standard unit interval. This leads to a notion of standard subsets of the natural numbers having the Standard Interval Measure Property and we enumerate some basic facts concerning sets with this property.

## 15.1 IM Sets

Let $I := [y, z]$ be an infinite, hyperfinite interval. Set $\mathrm{st}_I := \mathrm{st}_{[y,z]} : I \to [0, 1]$ to be the map $\mathrm{st}_I(a) := \mathrm{st}(\frac{a-y}{z-y})$. For $A \subseteq {}^*\mathbb{N}$ internal, we set $\mathrm{st}_I(A) := \mathrm{st}_I(A \cap I)$. We recall that $\mathrm{st}_I(A)$ is a closed subset of $[0, 1]$ and we may thus consider $\lambda_I(A) := \lambda(\mathrm{st}_I(A))$, where $\lambda$ is Lebesgue measure on $[0, 1]$.

We also consider the quantity $g_A(I) := \frac{d-c}{|I|}$, where $[c, d] \subseteq I$ is maximal so that $[c, d] \cap A = \emptyset$.

The main concern of this subsection is to compare the notions of making $g_A(I)$ small (an internal notion) and making $\lambda_I(A)$ large (an external notion). There is always a connection in one direction:

**Lemma 15.1** *If* $\lambda_I(A) > 1 - \epsilon$, *then* $g_A(I) < \epsilon$.

*Proof* Suppose that $g_A(I) \geq \epsilon$, whence there is $[c, d] \subseteq I$ such that $[c, d] \cap A = \emptyset$ and $\frac{d-c}{|I|} \geq \epsilon$. It follows that, for any $\delta > 0$, we have $(\mathrm{st}_I(c) + \delta, \mathrm{st}_I(d) - \delta) \cap$

© Springer Nature Switzerland AG 2019
M. Di Nasso et al., *Nonstandard Methods in Ramsey Theory and Combinatorial Number Theory*, Lecture Notes in Mathematics 2239, https://doi.org/10.1007/978-3-030-17956-4_15

$\text{st}_I(A) = \emptyset$, whence

$$\lambda_I(A) \leq 1 - \left( \text{st}\left( \frac{d-c}{|I|} \right) - 2\delta \right) \leq 1 - \epsilon + 2\delta.$$

Letting $\delta \to 0$ yields the desired result.

We now consider sets where there is also a relationship in the other direction.

**Definition 15.2** We say that $A$ has the *interval-measure property* (or *IM property*) on $I$ if for every $\epsilon > 0$, there is $\delta > 0$ such that, for all infinite $J \subseteq I$ with $g_A(J) \leq \delta$, we have $\lambda_J(A) \geq 1 - \epsilon$.

If $A$ has the IM property on $I$, we let $\delta(A, I, \epsilon)$ denote the supremum of the $\delta$'s that witness the conclusion of the definition for the given $\epsilon$.

It is clear from the definition that if $A$ has the IM property on an interval, then it has the IM property on every infinite subinterval. Also note that it is possible that $A$ has the IM property on $I$ for a trivial reason, namely that there is $\delta > 0$ such that $g_A(J) > \delta$ for every infinite $J \subseteq I$. Let us temporarily say that $A$ has the *nontrivial IM property* on $I$ if this does *not* happen, that is, for every $\delta > 0$, there is an infinite interval $J \subseteq I$ such that $g_A(J) \leq \delta$. It will be useful to reformulate this in different terms. In order to do that, we recall an important standard tool that is often employed in the study of sets with the IM property, namely the *Lebesgue density theorem*. Recall that for a measurable set $E \subseteq [0, 1]$, a point $r \in E$ is a *(one-sided) point of density of $E$* if

$$\lim_{s \to r^+} \frac{\mu(E \cap [r, s])}{s - r} = 1.$$

The Lebesgue density theorem asserts that almost every point of $E$ is a density point of $E$.

**Proposition 15.3** *Suppose that $A \subseteq {}^*\mathbb{N}$ is internal and $I$ is an infinite, hyperfinite interval such that $A$ has the IM property on $I$. Then the following are equivalent:*

1. *There is an infinite subinterval $J$ of $I$ such that $A$ has the nontrivial IM property on $J$.*
2. *There is an infinite subinterval $J$ of $I$ such that $\lambda_J(A) > 0$.*

*Proof* First suppose that $J$ is an infinite subinterval of $I$ such that $A$ has the nontrivial IM property on $J$. Let $J'$ be an infinite subinterval of $J$ such that $g_A(J') \leq \delta(A, J, \frac{1}{2})$. It follows that $\lambda_{J'}(A) \geq \frac{1}{2}$.

Now suppose that $J$ is an infinite subinterval of $I$ such that $\lambda_J(A) > 0$. By the Lebesgue density theorem, there is an infinite subinterval $J'$ of $J$ such that $\lambda_{J'}(A) > 1 - \delta$. By Lemma 15.1, we have that $g_A(J') < \delta$, whence $g_A(J) < \delta$. It follows that $A$ has the nontrivial IM property on $J$.

In practice, the latter property in the previous proposition is easier to work with. Consequently, let us say that $A$ has the *enhanced IM property on $I$* if it has the IM property on $I$ and $\lambda_I(A) > 0$.[1]

We now seek to establish nice properties of sets with the IM property. We first establish a kind of partition regularity theorem.

**Theorem 15.4** *Suppose that $A$ has the enhanced IM property on $I$. Further suppose that $A \cap I = B_1 \cup \cdots \cup B_n$ with each $B_i$ internal. Then there is $i$ and infinite $J \subseteq I$ such that $B_i$ has the enhanced IM property on $J$.*

*Proof* We prove the theorem by induction on $n$. The result is clear for $n = 1$. Now suppose that the result is true for $n - 1$ and suppose $A \cap I = B_1 \cup \cdots \cup B_n$ with each $B_i$ internal. If there is an $i$ and infinite $J \subseteq I$ such that $B_i \cap J = \emptyset$ and $\lambda_J(A) > 0$, then we are done by induction. We may thus assume that whenever $\lambda_J(A) > 0$, then each $B_i \cap J \neq \emptyset$. We claim that this implies that each of the $B_i$ have the IM property on $I$. Since there must be an $i$ such that $\lambda_I(B_i) > 0$, for such an $i$ it follows that $B_i$ has the enhanced IM property on $I$.

Fix $i$ and set $B := B_i$. Suppose that $J \subseteq I$ is infinite, $\epsilon > 0$, and $g_B(J) \leq \delta(A, I, \epsilon)$; we show that $\lambda_J(B) \geq 1 - \epsilon$. Since $g_A(J) \leq g_B(J) \leq \delta(A, I, \epsilon)$, we have that $\lambda_J(A) \geq 1 - \epsilon$. Suppose that $[r, s] \subseteq [0, 1] \setminus \mathrm{st}_J(B)$. Then $r = \mathrm{st}_J(x)$ and $s = \mathrm{st}_J(y)$ with $\frac{y-x}{|J|} \approx s - r$ and $B \cap [x, y] = \emptyset$. By our standing assumption, this implies that $\lambda_{[x,y]}(A) = 0$, whence it follows that $\lambda_J(A \cap [x, y]) = 0$. It follows that $\lambda_J(B) = \lambda_J(A) \geq 1 - \epsilon$, as desired.

If $A$ has the IM property on an interval $I$ and we have a subinterval of $I$ on which $A$ has small gap ratio, then by applying the IM property, the Lebesgue density theorem, and Lemma 15.1, we can find a smaller, but appreciably sized, subinterval on which $A$ once again has small gap ratio. Roughly speaking, one can iterate this procedure until one finds a *finite* subinterval of $I$ on which $A$ has small gap ratio; the finiteness of the subinterval will be crucial for applications. We now give a precise formulation.

Fix internal sets $A_1, \ldots, A_n$ and intervals $I_1, \ldots, I_n$. Fix also $\delta > 0$. A $\delta$-*configuration* (with respect to $A_1, \ldots, A_n, I_1, \ldots, I_n$) is a sequence of subintervals $J_1, \ldots, J_n$ of $I_1, \ldots, I_n$ respectively such that each $|J_i|$ has the same length and such that $g_{A_i}(J_i) \leq \delta$ for all $i$. We call the common length of the $J_i$'s the *length* of the configuration. There is an obvious notion of $\delta$-*subconfiguration*, although, for our purposes, we will need a stronger notion of subconfiguration. Indeed, we say that a $\delta$-subconfiguration $J_1', \ldots, J_n'$ of $J_1, \ldots, J_n$ is a *strong $\delta$-subconfiguration* if there is some $c \in {}^*\mathbb{N}$ such that, writing $a_i$ for the left endpoint of $J_i$, we have that $a_i + c$ is the left endpoint of $J_i'$. Note that the strong $\delta$-subconfiguration relation is transitive.

**Theorem 15.5** *Suppose that $A_1, \ldots, A_n$ are internal sets that satisfy the IM property on $I_1, \ldots, I_n$ respectively. Fix $\epsilon > 0$ such that $\epsilon < \frac{1}{n}$. Take $\delta > 0$ with*

---

[1] This terminology does not appear in the original article of Leth.

$\delta < \min_{i=1,\ldots,n} \delta(A_i, I_i, \epsilon)$. *Then there is $w \in \mathbb{N}$ such that any $\delta$-configuration has a strong $\delta$-subconfiguration of length at most $w$.*

*Proof* Let $A_1, \ldots, A_n, I_1, \ldots, I_n, \epsilon$ and $\delta$ be as in the statement of the theorem. The entire proof rests on the following:

*Claim* Any $\delta$-configuration of infinite length has a proper strong $\delta$-subconfiguration.

Given the claim, the proof of the theorem proceeds as follows: let $\mathscr{C}$ denote the internal set of $\delta$-configurations. Let $f : \mathscr{C} \to {}^*\mathbb{N}$ be the internal function given by $f(J_1, \ldots, J_n) =$ the minimal length of a minimal strong $\delta$-subconfiguration of $J_1, \ldots, J_n$. By the claim, the range of $f$ is contained in $\mathbb{N}$. Thus, there is $w \in \mathbb{N}$ such that the range of $f$ is contained in $[1, w]$, as desired.

Thus, to finish the proof of the theorem, it suffices to prove the claim.

*Proof of Claim* Write $J_i := [a_i, a_i + b]$ for $i = 1, \ldots, n$. By assumption, $\lambda_{J_i}(A_i) \geq 1 - \frac{1}{n}$, whence $\lambda(\bigcap_{i=1}^{n} \mathrm{st}_{J_i}(A_i)) > 0$. Let $r$ be a point of density for $\bigcap_{i=1}^{n} \mathrm{st}_{J_i}(A_i)$. Thus, there is $s < 1 - r$ such that

$$\lambda\left(\left(\bigcap_{i=1}^{n} \mathrm{st}_{J_i}(A_i)\right) \cap [r, r + s]\right) \geq (1 - \delta)s.$$

Set $c := \lfloor r \cdot b \rfloor$ and $b' := \lfloor s \cdot b \rfloor$. Then $c + b' \leq b$ and, by Lemma 15.1, we have

$$g_{A_i}([a_i + c, a_i + c + b']) \leq \delta \text{ for all } i = 1, \ldots, n.$$

Thus, the $[a_i + c, a_i + c + b']$ form the desired proper strong $\delta$-subconfiguration.

A special case of Theorem 15.5 is worth singling out:

**Corollary 15.6** *Let $A_1, \ldots, A_n, I_1, \ldots, I_n, \epsilon$, and $\delta$ be as in Theorem 15.5. Then there is $w \in \mathbb{N}$ such that, whenever $[a_i, a_i + b]$ is a $\delta$-configuration, then there is $c \in {}^*\mathbb{N}$ such that*

$$A_i \cap [a_i + c, a_i + c + w] \neq \emptyset \text{ for all } i = 1, \ldots, n.$$

By refining the proof of Theorem 15.5, we obtain the following:

**Corollary 15.7** *If $A$ has the IM property on $I$, then there is $w \in \mathbb{N}$ and a descending hyperfinite sequence $I = I_0, I_1, \ldots, I_K$ of hyperfinite subintervals of $I$ such that:*

- $|I_K| \leq w$;
- $\frac{|I_{k+1}|}{|I_k|} \geq \frac{1}{w}$;
- *whenever $I_k$ is infinite, we have $\lambda_{I_k}(A) > 0$.*

*Proof* First note that the proof of the Claim in Theorem 15.5 actually yields that every $\delta$-configuration has a proper strong $\delta$-subconfiguration where the ratio of lengths is non-infinitesimal. Thus, a saturation argument yields $\epsilon > 0$ such that every $\delta$-configuration has a proper strong $\delta$-subconfiguration with ratio of lengths at least $\epsilon$. The corollary follows easily from this, specializing to the case of a single internal set on a single interval.

**Definition 15.8** For any (not necessarily internal) $A \subseteq {}^*\mathbb{N}$, we set

$$D(A) := \{n \in \mathbb{N} : n = a - a' \text{ for infinitely many pairs } a, a' \in A\}.$$

The following corollary will be important for our standard application in the next section.

**Corollary 15.9** *Suppose that $A$ has the enhanced IM property on $I$. Then $D(A)$ is syndetic.*

*Proof* Let $w \in \mathbb{N}$ be as in Corollary 15.6 for $A_1 = A_2 = A$ and $I_1 = I_2 = I$. It suffices to show that for all $m \in \mathbb{N}$, there are infinitely many pairs $(x, y) \in A^2$ such that $y - x \in [m - w, m + w]$ (as then $[m - w, m + w] \cap D(A) \neq \emptyset$).

By considering countably many distinct points of density of $\mathrm{st}_I(A)$ and using Lemma 15.1 and overflow, we may find pairwise disjoint infinite subintervals $J_n := [a_n, b_n] \subseteq I$ such that $g_A(J_n) \approx 0$. Note also that $g_A(J_n + m) \approx 0$. Thus, by the choice of $w$, for each $n$, there is $c_n \in {}^*\mathbb{N}$ such that

$$A \cap [a_n + c_n, a_n + c_n + w], A \cap [a_n + m + c_n, a_n + m + c_n + w] \neq \emptyset.$$

If $x_n \in A \cap [a_n + c_n, a_n + c_n + w]$ and $y_n \in A \cap [a_n + m + c_n, a_n + m + c_n + w]$, then $y_n - x_n \in (A - A) \cap [m - w, m + w]$. By construction, the pairs $(x_n, y_n)$ are all distinct. $\blacksquare$

## 15.2 SIM Sets

We now seek to extract the standard content of the previous section.

**Definition 15.10** $A \subseteq \mathbb{N}$ has the *standard interval-measure property* (or *SIM property*) if:

- ${}^*A$ has the IM property on every infinite hyperfinite interval;
- ${}^*A$ has the enhanced IM property on some infinite hyperfinite interval.

*Example 15.11* Let $A = \bigcup_n I_n$, where each $I_n$ is an interval, $|I_n| \to \infty$ as $n \to \infty$, and there is $k \in \mathbb{N}$ such that the distance between consecutive $I_n$'s is at most $k$. Then $A$ has the SIM property.

We now reformulate the definition of SIM set using only standard notions. (Although recasting the SIM property in completely standard terms is not terribly illuminating, it is the polite thing to do.) First, note that one can define $g_A(I)$ for standard $A \subseteq \mathbb{N}$ and standard finite intervals $I \subseteq \mathbb{N}$ in the exact same manner. Now, for $A \subseteq \mathbb{N}$ and $0 < \delta < \epsilon < 1$, define the function $F_{\delta,\epsilon,A}$ :

$\mathbb{N} \to \mathbb{N}$ as follows. First, if $g_A(I) > \delta$ for every $I \subseteq \mathbb{N}$ of length $\geq n$, set $F_{\delta,\epsilon,A}(n) = 0$. Otherwise, set $F_{\delta,\epsilon,A}(n) =$ the minimum $k$ such that there is an interval $I \subseteq \mathbb{N}$ of length $\geq n$ such that $g_A(I) \leq \delta$ and there are subintervals $I_1, \ldots, I_k \subseteq I$ with $I_i \cap A = \emptyset$ for all $i = 1, \ldots, k$ and $\sum_{i=1}^k |I_i| \geq \epsilon |I|$.

**Theorem 15.12** *A has the SIM property if and only if: for all $\epsilon > 0$, there is $\delta > 0$ such that $\lim_{n \to \infty} F_{\delta,\epsilon,A}(n) = \infty$.*

*Proof* First suppose that there is $\epsilon > 0$ such that $\liminf_{n \to \infty} F_{\delta,\epsilon,A}(n) < \infty$ for all $\delta > 0$; we show that $A$ does not have the SIM property. Towards this end, we may suppose that $\lambda_I(^*A) > 0$ for some infinite hyperfinite interval $I$ and show that $^*A$ does not have the IM property on some infinite interval. Fix $0 < \delta < \epsilon$. By the Lebesgue density theorem and Lemma 15.1, we have that $g_{*A}(J) \leq \delta$ for some infinite subinterval $J \subseteq I$. By transfer, there are intervals $J_n \subseteq \mathbb{N}$ of length $\geq n$ such that $g_A(J_n) \leq \delta$, whence $0 < \liminf_{n \to \infty} F_{\delta,\epsilon,A}(n)$ for all $0 < \delta < \epsilon$. For every $k \geq 1$, set $m_k := 1 + \liminf_{n \to \infty} F_{\frac{1}{k},\epsilon,A}(n)$. Consequently, for every $n \in \mathbb{N}$, there are intervals $I_{1,n}, \ldots, I_{n,n}$ of length $\geq n$ such that, for each $k = 1, \ldots, n$, $g_A(I_{k,n}) \leq \frac{1}{k}$ and the sum of the lengths of $m_k$ many gaps of $A$ in $I_{k,n}$ is at least $\epsilon \cdot |I_{k,n}|$. Set $I_n := I_{1,n} \cup \cdots \cup I_{n,n}$. By overflow, there is an infinite, hyperfinite interval $\tilde{I}$ that contains infinite subintervals $I_k$ such that $g_{*A}(I_k) \leq \frac{1}{k}$ and yet the sum the lengths of $m_k$ many gaps of $A$ on $I_k$ have size at least $\epsilon|I_k|$. It follows that $^*A$ does not have the IM property on $\tilde{I}$.

Now suppose that for all $\epsilon > 0$, there is $\delta > 0$ such that $\lim_{n \to \infty} F_{\delta,\epsilon,A}(n) = \infty$ and that $I$ is an infinite, hyperfinite interval such that $g_{*A}(I) \leq \delta$. By transfer, it follows that no finite number of gaps of $^*A$ on $I$ have size at least $\epsilon \cdot |I|$. Since $\mathrm{st}_I(^*A)$ is closed, we have that $\lambda_I(^*A) \geq 1 - \epsilon$. Consequently, $A$ has the IM property on any infinite, hyperfinite interval. Since, by transfer, there is an infinite, hyperfinite interval $I$ with $g_{*A}(I) \leq \delta$, this also shows that $^*A$ has the enhanced IM property on this $I$. Consequently, $A$ has the SIM property. $\qquad \square$

**Exercise 15.13** Suppose that $A \subseteq \mathbb{N}$ has the SIM property. Show that, for each $\epsilon > 0$, there is a $\delta > 0$ such that $\delta \leq \delta(^*A, I, \epsilon)$ *for every* infinite hyperfinite intervals $I \subseteq {}^*\mathbb{N}$.

The next lemma shows that the SIM property is not simply a measure of "largeness" as this property is not preserved by taking supersets.

**Lemma 15.14** *Suppose that $A \subseteq \mathbb{N}$ is not syndetic. Then there is $B \supseteq A$ such that $B$ does not have the SIM property.*

*Proof* For each $n$, let $x_n \in \mathbb{N}$ be such that $[x_n, x_n + n^2] \cap A = \emptyset$. Let

$$B := A \cup \bigcup_n \{x_n + kn : k = 0, 1, \ldots n\}.$$

Fix $\epsilon > 0$. Take $m \in \mathbb{N}$ such that $m > \frac{1}{\epsilon}$ and take $N > \mathbb{N}$. Set $I := [x_N, x_N + mN]$. Indeed, $g_{*B}(I) = \frac{N}{mN} < \frac{1}{\epsilon}$ while

$$\mathrm{st}_I(^*B) = \left\{ \mathrm{st}\left( \frac{kN}{mN} \right) \ : \ k = 0, \ldots, m \right\} = \left\{ \frac{k}{m} \ : \ k = 0, \ldots, m \right\}$$

is finite and thus has measure 0. It follows that $^*B$ does not have the IM property on $I$, whence $B$ does not have the SIM property.

The previous lemma also demonstrates that one should seek structural properties of a set which ensure that it contains a set with the SIM property. Here is an example:

**Lemma 15.15** *If $B$ is piecewise syndetic, then there is $A \subseteq B$ with the SIM property.*

*Proof* For simplicity, assume that $B$ is thick; the argument in general is similar, just notationally more messy. Let $A := \bigcup_n I_n$, with $I_n$ intervals contained in $B$, $|I_n| \to \infty$ as $n \to \infty$, and such that, setting $g_n$ to be the length in between $I_n$ and $I_{n+1}$, we have $g_{n+1} \geq n g_n$ for all $n$. We claim that $A$ has the SIM property. It is clear that $\lambda_I(^*A) > 0$ for some infinite hyperfinite interval $I$; indeed, $\lambda_{I_N}(^*A) = 1$ for $N > \mathbb{N}$. Now suppose that $I$ is an infinite hyperfinite interval; we claim that $^*A$ has the IM property on $I$ as witnessed by $\delta = \epsilon$. Suppose that $J$ is an infinite subinterval of $I$ such that $g_{*A}(J) \leq \epsilon$. Suppose that $I_n, \ldots, I_{M+1}$ is a maximal collection of intervals from $^*A$ intersecting $J$. Since $\frac{g_M}{|J|} \leq \epsilon$, for $k = N, \ldots, M-1$, we have $\frac{g_k}{|J|} = \frac{g_k}{g_M} \cdot \frac{g_M}{|J|} \approx 0$, whence the intervals $I_n, \ldots, I_M$ merge when one applies $\mathrm{st}_J$. It follows that $\lambda_J(^*A) \geq 1 - \epsilon$.

In connection with the previous result, the following question seems to be the most lingering open question about sets that contain subsets with the SIM property:

*Question 15.16* Does every set of positive Banach density contain a subset with the SIM property?

The next result shows that many sets do *not* have the SIM property.

**Proposition 15.17** *Suppose that $A = (a_n)$ is a subset of $\mathbb{N}$ written in increasing order. Suppose that $\lim_{n \to \infty} (a_{n+1} - a_n) = \infty$. Then $A$ does not have the SIM property.*

*Proof* Suppose that $A$ has the SIM property. Take $I$ such that $\lambda_I(^*A) > 0$. Then by the proof of Corollary 15.9, we can find $x, y \in {}^*A \backslash A$ such that $x < y$ and $y - x \leq 2w$. Then, by transfer, there are arbitrarily large $m, n \in \mathbb{N}$ with $0 < m - n \leq 2w$. It follows that $\lim_{n \to \infty} (a_{n+1} - a_n) \neq \infty$.

The following theorem provides a connection between the current chapter and the previous one. The proof follows immediately from Theorem 15.5.

**Theorem 15.18** *Suppose that $A \subseteq \mathbb{N}$ is a SIM set. Fix $t \in \mathbb{N}$ and $0 < \epsilon < \frac{1}{t}$. Let $\delta > 0$ be as in Exercise 15.13 for $\epsilon$. Then there is $j \in \mathbb{N}$ such that whenever*

*A nearly contains a* $(t, d, w)$-*progression* $B(b, t, d, w)$, *then A nearly contains a* subprogression[2] $B(b', t, d, j)$ *of* $B(b, t, d, w)$.

We end this section with a result concerning a structural property of sets with the SIM property. A direct consequence of Corollary 15.9 is the following:

**Corollary 15.19** *If A has the SIM property, then* $D(A)$ *is syndetic.*

Leth's original main motivation for studying the IM property was a generalization of the previous corollary. Stewart and Tijdeman [118] proved that, given $A_1, \ldots, A_n \subseteq \mathbb{N}$ with $BD(A_i) > 0$ for all $i = 1, \ldots, n$, one has $D(A_1) \cap \cdots \cap D(A_n)$ is syndetic. Leth proved the corresponding statement for sets with the SIM property:

**Theorem 15.20** *If* $A_1, \ldots, A_n \subseteq \mathbb{N}$ *all have the SIM property, then* $D(A_1) \cap \cdots \cap D(A_n)$ *is syndetic.*

*Proof* We break the proof up into pieces.

*Claim 1* There are infinite hyperfinite intervals $I_1, \ldots, I_n$, all of which have the same length, such that $\lambda_{I_i}(^*A_i) = 1$ for all $i = 1, \ldots, n$.

*Proof of Claim 1* By the definition of the SIM property and Corollary 15.7, we may find infinite, hyperfinite intervals $J_1, \ldots, J_n$ whose length ratios are all finite and for which $\mathrm{st}_{J_i}(^*A_i) > 0$ for $i = 1, \ldots, n$. By taking points of density in each of these intervals, for any $\epsilon > 0$, we may find *equally sized* subintervals $J_i'$ of $J_i$ such that $\lambda_{J_i'}(^*A_i) \geq 1 - \epsilon$, whence $g_{*A_i}(J_i') \leq \epsilon$. Since this latter condition is internal, by saturation, we may find equally sized subintervals $I_i$ of $J_i$ such that each $g_{*A_i}(I_i) \approx 0$, whence, by the fact that $^*A_i$ has the IM property on $J_i$, we have $\lambda_{I_i}(^*A_i) = 1$.

We now apply Corollary 15.6 to $A_1, \ldots, A_n, I_1, \ldots, I_n$ and $\epsilon := \frac{1}{n+1}$. Let $w \in \mathbb{N}$ be as in the conclusion of that corollary. Write $I_i := [x_i, y_i]$ and for $i = 1, \ldots, n$, set $d_i := x_i - x_1$. We then set

$$B := \{a \in \,^*A_1 \cap I_1 \,:\, ^*A_i \cap [a + d_i - w, a + d_i + 2w] \neq \emptyset \text{ for all } i = 1, \ldots, n\}.$$

*Claim 2* Suppose that $J \subseteq I_1$ is infinite and $r$ is a point of density of

$$\bigcap_{i=1}^{n} \mathrm{st}_{J+d_i}(^*A_i).$$

Then $r \in \mathrm{st}_J(B)$.

---

[2]Here, subprogression means that every block $[b'+id, b'+id+j]$ is contained in the corresponding block $[b + id, b + id + w]$.

*Proof of Claim 2* By a (hopefully) by now familiar Lebesgue density and overflow argument, there is an infinite hyperfinite interval $[u, v] \subseteq J$ such that $\mathrm{st}_J(u) = \mathrm{st}_J(v) = r$ and

$$g_{*A_i}([u + d_i, v + d_i]) \approx 0 \text{ for all } i = 1, \ldots, n.$$

This allows us to find $c \in {}^*\mathbb{N}$ such that $u + d_i + c + w \leq v_i$ and ${}^*A_i \cap [u + d_i + c, u + d_i + c + w] \neq \emptyset$ for $i = 1, \ldots, n$. Take $a \in {}^*A_1 \cap [u + c, u + c + w]$, say $a = u + c + j$ for $j \in [0, w]$. It follows that

$${}^*A_i \cap [a + d_i - j, a + d_i + j + w] \neq \emptyset \text{ for all } i = 1, \ldots, n$$

whence $a \in B$. Since $u \leq u + c \leq a \leq u + c + w \leq v$, we have that $\mathrm{st}_J(a) = \mathrm{st}_J(v) = r$, whence $r \in \mathrm{st}_J(B)$, as desired.

*Claim 3*  $B$ has the enhanced IM property on $I_1$.

*Proof of Claim 3* Taking $J = I_1$ in Claim 2 shows that $\lambda_{I_1}(B) = 1$. We now show that $B$ has the IM property on $I_1$. Fix $\epsilon > 0$. Let $\delta = \min_{i=1,\ldots,n} \delta({}^*A_i, I_i, \frac{\epsilon}{n})$. Suppose $J \subseteq I_1$ is such that $g_B(J) \leq \delta$. Then $g_{*A_i}(J + d_i) \leq \delta$, whence

$$\lambda\left(\bigcap_{i=1}^n \mathrm{st}_{J+d_i}({}^*A_i)\right) \geq 1 - \epsilon.$$

By Claim 2, we have $\lambda_J(B) \geq 1 - \epsilon$, as desired.

For $-w \leq k_1, \ldots, k_n \leq 2w$, set

$$B_{(k_1,\ldots,k_n)} := \{b \in B \; : \; b + d_i + k_i \in {}^*A_i \text{ for all } i = 1, \ldots, n\}.$$

By the definition of $B$, we have that $B$ is the union of these sets. Since $B$ has the enhanced IM property on $I_1$, by Theorem 15.4, there is such a tuple $(k_1, \ldots, k_n)$ and an infinite $J \subseteq I_1$ such that $B' := B_{(k_1,\ldots,k_n)}$ has the enhanced IM property on $J$. By Corollary 15.9, $D(B')$ is syndetic. Since $B' - B' \subseteq \bigcap_{i=1}^n({}^*A_i - {}^*A_i)$, by transfer we have that $D(A_1) \cap \cdots \cap D(A_n)$ is syndetic.

## Notes and References

The material in this chapter comes from the paper [88], although many of the proofs appearing above were communicated to us by Leth and are simpler than those appearing in the aforementioned article. In a recent preprint [55], Goldbring and Leth study the notion of a *supra-SIM* set, which is simply a set that contains a SIM set. They show that these sets have very nice properties such as being partition regular and closed under finite-embeddability. They also show that SIM sets satisfy the conclusions of the Sumset Theorem 12.7 and Nathansons' Theorem 13.4.

# Part IV
# Other Topics

# Chapter 16
# Triangle Removal and Szemerédi Regularity

In this chapter, we give nonstandard proofs of two of the more prominent results in extremal graph theory, namely the Triangle Removal Lemma and the Szemerédi Regularity Lemma.

## 16.1 Triangle Removal Lemma

The material in this section was not proven first by nonstandard methods. However, the nonstandard perspective makes the proofs quite elegant. We closely follow [123].

Suppose that $G = (V, E)$ is a finite graph. We define the *edge density* of $G$ to be the quantity

$$e(G) := \frac{|E|}{|V \times V|}$$

and the *triangle density* of $G$ to be the quantity

$$t(G) := \frac{|\{(x, y, z) \in V \times V \times V \ : \ (x, y), (y, z), (x, z) \in E\}|}{|V \times V \times V|}.$$

**Theorem 16.1 (Triangle Removal Lemma)** *For every $\epsilon > 0$, there is a $\delta > 0$ such that, whenever $G = (V, E)$ is a finite graph with $t(G) \leq \delta$, then there is a subgraph $G' = (V, E')$ of $G$ that is triangle-free (so $t(G') = 0$) and such that $e(G \backslash G') \leq \epsilon$.*

© Springer Nature Switzerland AG 2019
M. Di Nasso et al., *Nonstandard Methods in Ramsey Theory and Combinatorial Number Theory*, Lecture Notes in Mathematics 2239,
https://doi.org/10.1007/978-3-030-17956-4_16

In short, the triangle removal lemma says that if the triangle density of a graph is small, then one can remove a few number of edges to get one that is actually triangle-free. We first show how the Triangle Removal Lemma can be used to prove Roth's theorem, which was a precursor to Szemerédi's theorem.

**Theorem 16.2 (Roth's Theorem)**  *For all $\epsilon > 0$, there is $n_0 \in \mathbb{N}$ such that, for all $n \geq n_0$ and all $A \subseteq [1, n]$, if $\delta(A, n) \geq \epsilon$, then $A$ contains a 3-term arithmetic progression.*

*Proof* Fix $n$ and form a tripartite graph $G = G(A, n)$ with vertex set $V = V_1 \cup V_2 \cup V_3$, where each $V_i$ is a disjoint copy of $[1, 3n]$. If $(v, w) \in (V_1 \times V_2) \cup (V_2 \times V_3)$, we declare $(v, w) \in E \Leftrightarrow w - v \in A$. If $(v, w) \in V_1 \times V_3$, then we declare $(v, w) \in E \Leftrightarrow (w - v) \in 2A$. Note then that if $(v_1, v_2, v_3)$ is a triangle in $G$, then setting $a := v_2 - v_1$, $b := v_3 - v_2$, and $c := \frac{1}{2}(v_3 - v_1)$, we have that $a, b, c \in A$ and $a - c = c - b$. If this latter quantity is nonzero, then $\{a, b, c\}$ forms a 3-term arithmetic progression in $A$.

Motivated by the discussion in the previous paragraph, let us call a triangle $\{v_1, v_2, v_3\}$ in $G$ *trivial* if $v_2 - v_1 = v_3 - v_2 = \frac{1}{2}(v_3 - v_1)$. Thus, we aim to show that, for $n$ sufficiently large, if $\delta(A, n) \geq \epsilon$, then $G(A, n)$ has a nontrivial triangle. If $a \in A$ and $k \in [1, n]$, then $(k, k + a, k + 2a)$ is a trivial triangle in $G$. Since trivial triangles clearly do not share any edges, one would have to remove at least $3 \cdot |A| \cdot n \geq 3\epsilon n^2$ many edges of $G$ in order to obtain a triangle-free subgraph of $G$. Thus, if $\delta > 0$ corresponds to $3\epsilon$ in the triangle removal lemma, then we can conclude that $t(G) \geq \delta$, that is, there are at least $27\delta n^3$ many triangles in $G$. Since the number of trivial triangles is at most $|A| \cdot (3n) \leq 3n^2$, we see that $G$ must have a nontrivial triangle if $n$ is sufficiently large.

We now turn to the proof of the triangle removal lemma. The basic idea is that if the triangle removal lemma were false, then by a now familiar compactness/overflow argument, we will get a contradiction to some nonstandard triangle removal lemma. Here is the precise version of such a lemma:

**Theorem 16.3 (Nonstandard Triangle Removal Lemma)**  *Suppose that $V$ is a nonempty hyperfinite set and $E_{12}, E_{23}, E_{13} \in \mathscr{L}_{V \times V}$ are such that*

$$\int_{V \times V \times V} 1_{E_{12}}(u, v) 1_{E_{23}}(v, w) 1_{E_{13}}(u, v) d\mu(u, v, w) = 0. \quad (\dagger)$$

*Then for every $\epsilon > 0$ and $(i, j) \in \{(1, 2), (2, 3), (1, 3)\}$, there are hyperfinite $F_{ij} \subseteq V \times V$ such that $\mu_{V \times V}(E_{ij} \backslash F_{ij}) < \epsilon$ and*

$$1_{F_{12}}(u, v) 1_{F_{23}}(v, w) 1_{F_{13}}(u, v) = 0 \text{ for all } (u, v, w) \in V \times V \times V. \quad (\dagger\dagger)$$

**Proposition 16.4**  *The nonstandard triangle removal lemma implies the triangle removal lemma.*

*Proof* Suppose that the triangle removal lemma is false. Then there is $\epsilon > 0$ such that, for all $n \in \mathbb{N}$, there is a finite graph $G_n = (V_n, E_n)$ for which $t(G_n) \leq \frac{1}{n}$ and yet there does not exist a triangle-free subgraph $G' = (V_n, E'_n)$ with $|E_n \backslash E'_n| \leq \epsilon |V_n|^2$. Note that it follows that $|V_n| \to \infty$ as $n \to \infty$. By , there is an infinite hyperfinite graph $G = (V, E)$ such that $t(G) \approx 0$, whence (†) holds, and yet there does not exist a triangle-free hyperfinite subgraph $G' = (V, E')$ with $|E \backslash E'| \leq \epsilon |V|^2$. We claim that this latter statement yields a counterexample to the nonstandard triangle removal lemma. Indeed, if the nonstandard triangle removal held, then there would be hyperfinite $F_{ij} \subseteq V \times V$ such that $\mu_{V \times V}(E \backslash F_{ij}) < \frac{\epsilon}{6}$ and for which (††) held. If one then sets $E' := E \cap \bigcap_{ij}(F_{ij} \cap F_{ij}^{-1})$, then $G' = (V, E')$ is a hyperfinite subgraph of $G$ that is triangle-free and $\mu(E \backslash E') < \epsilon$, yielding the desired contradiction.[1]

It might look like the nonstandard triangle removal lemma is stated in a level of generality that is more than what is needed for we have $E_{12} = E_{23} = E_{13} = E$. However, in the course of proving the lemma, we will come to appreciate this added level of generality of the statement.

**Lemma 16.5** *Suppose that $f \in L^2(\mathcal{L}_{V \times V})$ is orthogonal to $L^2(\mathcal{L}_V \otimes \mathcal{L}_V)$. Then for any $g, h \in L^2(\mathcal{L}_{V \times V})$, we have*

$$\int_{V \times V \times V} f(x, y)g(y, z)h(x, z)d\mu_{V \times V \times V}(x, y, z) = 0.$$

*Proof* Fix $z \in V$. Let $g_z : V \to \mathbb{R}$ be given by $g_z(y) := g(y, z)$. Likewise, define $h_z(x) := h(x, z)$. Note then that $g_z \cdot h_z \in L^2(\mathcal{L}_V \otimes \mathcal{L}_V)$. It follows that

$$\int_{V \times V} f(x, y)g(y, z)h(x, z)d\mu_{V \times V}(x, y)$$

$$= \int_{V \times V} f(x, y)g_z(y)h_z(x)d\mu_{V \times V}(x, y) = 0.$$

By Theorem 5.21, we have that

$$\int_{V \times V \times V} f(x, y)g(y, z)h(x, z)d\mu_{V \times V \times V}(x, y, z)$$

$$= \int_V \left[ \int_{V \times V} f(x, y)g(y, z)h(x, z)d\mu_{V \times V}(x, y) \right] d\mu_V(z) = 0.$$

---

[1]Given a binary relation $R$ on a set $X$, we write $R^{-1}$ for the binary relation on $X$ given by $(x, y) \in R^{-1}$ if and only if $(y, x) \in R$.

*Proof of Theorem 16.3* We first show that we can assume that each $E_{ij}$ belongs to $\mathscr{L}_V \otimes \mathscr{L}_V$. Indeed, let $f_{ij} := \mathbb{E}[1_{E_{ij}} | \mathscr{L}_V \otimes \mathscr{L}_V]$.[2] Then by three applications of the previous lemma, we have

$$\int_{V \times V \times V} f_{12} f_{23} f_{13} d\mu_{V \times V \times V} = \int_{V \times V \times V} f_{12} f_{23} 1_{13} d\mu_{V \times V \times V}$$

$$= \int_{V \times V \times V} f_{12} 1_{23} 1_{13} d\mu_{V \times V \times V}$$

$$= \int_{V \times V \times V} 1_{12} 1_{23} 1_{13} d\mu_{V \times V \times V} = 0. \quad (*)$$

Let $G_{ij} := \{(u, v) \in V \times V : f_{ij}(u, v) \geq \frac{\epsilon}{3}\}$. Observe that each $G_{ij}$ belongs to $\mathscr{L}_V \otimes \mathscr{L}_V$ and

$$\mu(E_{ij} \backslash G_{ij}) = \int_{V \times V} 1_{E_{ij}} (1 - 1_{G_{ij}}) d\mu_{V \times V} = \int_{V \times V} f_{ij} (1 - 1_{G_{ij}}) d\mu_{V \times V} \leq \frac{\epsilon}{2}.$$

By $(*)$ we have

$$\int_{V \times V} 1_{G_{12}} 1_{G_{23}} 1_{G_{13}} d\mu_{V \times V} = 0.$$

Thus, if the nonstandard triangle removal lemma is true for sets belonging to $\mathscr{L}_V \otimes \mathscr{L}_V$, we can find hyperfinite $F_{ij} \subseteq V \times V$ such that $\mu(G_{ij} \backslash F_{ij}) < \frac{\epsilon}{2}$ and such that (††) holds. Since $\mu(E_{ij} \backslash F_{ij}) < \epsilon$, the $F_{ij}$ are as desired.

Thus, we may now assume that each $E_{ij}$ belongs to $\mathscr{L}_V \otimes \mathscr{L}_V$. Consequently, there are elementary sets $H_{ij}$ such that $\mu(E_{ij} \triangle H_{ij}) < \frac{\epsilon}{6}$. By considering the boolean algebra generated by the sides of the boxes appearing in the description of $H_{ij}$, we obtain a partition $V = V_1 \sqcup \cdots \sqcup V_n$ of $V$ into finitely many hyperfinite subsets of $V$ such that each $H_{ij}$ is a union of boxes of the form $V_k \times V_l$ for $k, l \in \{1, \ldots, n\}$. Let

$$F_{ij} := \bigcup \{V_k \times V_l : V_k \times V_l \subseteq H_{ij}, \ \mu(V_k \times V_l) > 0,$$

$$\text{and } \mu(E_{ij} \cap (V_k \times V_l)) > \frac{2}{3} \mu(V_k \times V_l)\}.$$

Clearly each $F_{ij}$ is hyperfinite. Note that

$$\mu(H_{ij} \backslash F_{ij}) = \mu((H_{ij} \backslash F_{ij}) \cap E_{ij}) + \mu((H_{ij} \backslash F_{ij}) \backslash E_{ij}) \leq \frac{2}{3} \mu(H_{ij} \backslash F_{ij}) + \frac{\epsilon}{6},$$

---

[2]Here, for $f \in L^2(\mathscr{L}_{V \times V})$, $\mathbb{E}[f | \mathscr{L}_V \otimes \mathscr{L}_V]$ denotes the conditional expectation of $f$ onto the subspace $L^2(\mathscr{L}_V \otimes \mathscr{L}_V)$.

whence $\mu(H_{ij} \setminus F_{ij}) \leq \frac{\epsilon}{2}$ and thus $\mu(E_{ij} \setminus F_{ij}) \leq \frac{\epsilon}{6} + \frac{\epsilon}{2} < \epsilon$. It remains to show that (††) holds. Towards a contradiction, suppose that $(u, v, w)$ witnesses that (††) is false. Take $k, l, m \in \{1, \ldots, n\}$ such that $u \in V_k$, $v \in V_l$, and $w \in V_m$. Since $(u, v) \in F_{12}$, we have that $\mu(E_{12} \cap (V_k \times V_l)) > \frac{2}{3}\mu(V_k \times V_l)$. Consequently, $\mu(E_{12} \times V_m) > \frac{2}{3}\mu(V_k \times V_l \times V_m)$. Similarly, we have that $\mu(E_{23} \times V_k)$, $\mu(E_{13} \times V_l) > \frac{2}{3}\mu(V_k \times V_l \times V_m)$. Thus, by elementary probability considerations, it follows that

$$\int_{V \times V} 1_{E_{12}} 1_{E_{23}} 1_{E_{13}} d\mu_{V \times V} > 0,$$

contradicting (†).

## 16.2 Szemerédi Regularity Lemma

Suppose that $(V, E)$ is a finite graph. For two nonempty subsets $X, Y$ of $V$, we define the *density of arrows between $X$ and $Y$* to be the quantity

$$d(X, Y) := \delta(E, X \times Y) = \frac{|E \cap (X \times Y)|}{|X||Y|}.$$

For example, if every element of $X$ is connected to every element of $Y$ by an edge, then $d(X, Y) = 1$. Fix $\epsilon \in \mathbb{R}^{>0}$. We say that $X$ and $Y$ as above are $\epsilon$-*pseudorandom* if whenever $A \subseteq X$ and $B \subseteq Y$ are such that $|A| \geq \epsilon|X|$ and $|B| \geq \epsilon|Y|$, then $|d(A, B) - d(X, Y)| < \epsilon$. In other words, as long as $A$ and $B$ contain at least an $\epsilon$ proportion of the elements of $X$ and $Y$ respectively, then $d(A, B)$ is essentially the same as $d(X, Y)$, so the edges between $X$ and $Y$ are distributed in a sort of random fashion.

If $X = \{x\}$ and $Y = \{y\}$ are singletons, then clearly $X$ and $Y$ are $\epsilon$-pseudorandom for any $\epsilon$. Thus, any finite graph can trivially be partitioned into a finite number of $\epsilon$-pseudorandom pairs by partitioning the graph into singletons. Szemerédi's Regularity Lemma essentially says that one can do much better in the sense that there is a constant $C(\epsilon)$ such that any finite graph has an "$\epsilon$-pseudorandom partition" into at most $C(\epsilon)$ pieces. Unfortunately, the previous sentence is not entirely accurate as there is a bit of error that we need to account for.

Suppose that $V_1, \ldots, V_m$ is a partition of $V$ into $m$ pieces. Set

$$R := \{(i, j) \mid 1 \leq i, j \leq m, \ V_i \text{ and } V_j \text{ are } \epsilon\text{-pseudorandom}\}.$$

We say that the partition is $\epsilon$-*regular* if $\sum_{(i,j)\in R} \frac{|V_i||V_j|}{|V|^2} > (1-\epsilon)$. This says that, in some sense, almost all of the pairs of points are in $\epsilon$-pseudorandom pairs. We can now state:

**Theorem 16.6 (Szemerédi's Regularity Lemma)** *For any $\epsilon \in \mathbb{R}^{>0}$, there is a constant $C(\epsilon)$ such that any graph $(V, E)$ admits an $\epsilon$-regular partition into $m \leq C(\epsilon)$ pieces.*

As in the previous section, the regularity lemma is equivalent to a nonstandard version of the lemma. We leave the proof of the equivalence as an exercise to the reader.

**Proposition 16.7** *Szemerédi's Regularity Lemma is equivalent to the following statement: for any $\epsilon$ and any hyperfinite graph $(V, E)$, there is a finite partition $V_1, \ldots, V_m$ of $V$ into internal sets and a subset $R \subseteq \{1, \ldots, m\}^2$ such that:*

- *for $(i, j) \in R$, $V_i$ and $V_j$ are internally $\epsilon$-pseudorandom: for all internal $A \subseteq V_i$ and $B \subseteq V_j$ with $|A| \geq \epsilon|V_i|$ and $|B| \geq \epsilon|V_j|$, we have $|d(A, B) - d(V_i, V_j)| < \epsilon$; and*
- $\sum_{(i,j)\in R} \frac{|V_i||V_j|}{|V|^2} > (1-\epsilon)$.

We will now prove the above nonstandard equivalent of the Szemerédi Regularity Lemma. Fix $\epsilon$ and a hyperfinite graph $(V, E)$. Set $f := \mathbb{E}[1_E | \mathscr{L}_V \otimes \mathscr{L}_V]$. The following calculation will prove useful: Suppose that $A, B \subseteq V$ are internal and $\frac{|A|}{|V|}$ and $\frac{|B|}{|V|}$ are noninfinitesimal. Then ($\clubsuit$):

$$\int_{A\times B} f d(\mu_V \otimes \mu_V) = \int_{A\times B} 1_E d\mu_{V\times V} \quad \text{by the definition of } f$$

$$= \text{st}\left(\frac{|E \cap (A \times B)|}{|V|^2}\right)$$

$$= \text{st}\left(\frac{|E \cap (A \times B)|}{|A||B|}\right) \text{st}\left(\frac{|A||B|}{|V|^2}\right)$$

$$= \text{st}(d(A, B)) \text{st}\left(\frac{|A||B|}{|V|^2}\right).$$

Fix $r \in \mathbb{R}^{>0}$, to be determined later. Now, since $f$ is $\mu_V \otimes \mu_V$-integrable, there is a $\mu_V \otimes \mu_V$-simple function $g \leq f$ such that $\int (f - g)d(\mu_V \otimes \mu_V) < r$. Set $C := \{\omega \in V \times V \mid f(\omega) - g(\omega) \geq \sqrt{r}\} \in s_V \otimes s_V$. Then $(\mu_V \otimes \mu_V)(C) < \sqrt{r}$, for otherwise

$$\int (f-g)d(\mu_V\otimes\mu_V) \geq \int_C (f-g)d(\mu_V\otimes\mu_V) \geq \int_C \sqrt{r}d(\mu_V\otimes\mu_V) \geq \sqrt{r}\sqrt{r} = r.$$

By Fact 5.10, there is an elementary set $D \in s_V \otimes s_V$ that is a finite, disjoint union of rectangles of the form $V' \times V''$, with $V'$, $V'' \subseteq V$ internal sets, such that $C \subseteq D$ and $(\mu_V \otimes \mu_V)(D) < \sqrt{r}$. In a similar way, we may assume that the level sets of $g$ (that is, the sets on which $g$ takes constant values) are elementary sets (Exercise). We now take a finite partition $V_1, \ldots, V_m$ of $V$ into internal sets such that $g$ and $1_D$ are constant on each rectangle $V_i \times V_j$. For ease of notation, set $d_{ij}$ to be the constant value of $g$ on $V_i \times V_j$.

*Claim* If $\mu_V(V_i)$, $\mu_V(V_j) \neq 0$ and $(V_i \times V_j) \cap D = \emptyset$, then $V_i$ and $V_j$ are internally $2\sqrt{r}$-pseudorandom.

*Proof of Claim* Since $C \subseteq D$, we have that $(V_i \times V_j) \cap C = \emptyset$, whence

$$d_{ij} \leq f(\omega) < d_{ij} + \sqrt{r} \text{ for } \omega \in V_i \times V_j. \quad (\clubsuit\clubsuit).$$

Now suppose that $A \subseteq V_i$ and $B \subseteq V_j$ are such that $|A| \geq 2\sqrt{r}|V_i|$ and $|B| \geq 2\sqrt{r}|V_j|$. In particular, $\frac{|A|}{|V_i|}$ and $\frac{|B|}{|V_j|}$ are noninfinitesimal. Since $\mu_V(V_i), \mu_V(V_j) > 0$, it follows that $\frac{|A|}{|V|}$ and $\frac{|B|}{|V|}$ are noninfinitesimal and the calculation $(\clubsuit)$ applies. Integrating the inequalities $(\clubsuit\clubsuit)$ on $A \times B$ yields:

$$d_{ij} \operatorname{st}\left(\frac{|A||B|}{|V|^2}\right) \leq \operatorname{st}(d(A, B)) \operatorname{st}\left(\frac{|A||B|}{|V|^2}\right) < (d_{ij} + \sqrt{r}) \operatorname{st}\left(\frac{|A||B|}{|V|^2}\right).$$

We thus get:

$$|d(A, B) - d(V_i, V_j)| \leq |d(A, B) - d_{ij}| + |d(V_i, V_j) - d_{ij}| < 2\sqrt{r}.$$

By the Claim, we see that we should choose $r < (\frac{\epsilon}{2})^2$, so $V_i$ and $V_j$ are internally $\epsilon$-pseudorandom when $V_i$ and $V_j$ are non-null and satisfy $(V_i \times V_j) \cap D = \emptyset$. It remains to observe that the $\epsilon$-pseudorandom pairs almost cover all pairs of vertices. Let $R := \{(i, j) \mid V_i \text{ and } V_j \text{ are } \epsilon\text{-pseudorandom}\}$. Then

$$\operatorname{st}\left(\sum_{(i,j) \in R} \frac{|V_i||V_j|}{|V|^2}\right) = \mu_{V \times V}\left(\bigcup_{(i,j) \in R} (V_i \times V_j)\right)$$

$$\geq \mu_{V \times V}((V \times V) \backslash D)$$

$$> 1 - \sqrt{r}$$

$$> 1 - \epsilon.$$

This finishes the proof of the Claim and the proof of the Szemerédi Regularity Lemma.

## Notes and References

The triangle removal lemma was originally proven by Ruzsa and Szemerédi in [113] and their proof used several applications of an early version of the Szemerédi regularity lemma. Nowadays, the most common standard proof of the triangle removal lemma goes through a combination of the Szemerédi regularity lemma and the so-called *Counting lemma*; see, for example, [58]. Szemerédi's regularity lemma was a key ingredient in his proof in [119] that sets of positive density contain arbitrarily long arithmetic progressions (which we now of course call Szemerédi's theorem). Analogous to the above proof of Roth's Theorem from the Triangle Removal Lemma, one can prove Szemeredi's theorem by first proving an appropriate removal lemma called the *Hypergraph removal lemma* and then coding arithmetic progressions by an appropriate hypergraph generalization of the argument given above. For more details, see [58] for the original standard proof and [120] and [56] for simplified nonstandard proofs.

# Chapter 17
# Approximate Groups

In this chapter, we describe a recent application of nonstandard methods to multiplicative combinatorics, namely to the structure theorem for finite approximate groups. The general story is much more complicated than the rest of the material in this book and there are already several good sources for the complete story (see [20] or [128]), so we content ourselves to a summary of some of the main ideas. Our presentation will be similar to the presentation from [128].

## 17.1 Statement of Definitions and the Main Theorem

In this chapter, $(G, \cdot)$ denotes an arbitrary group and $K \in \mathbb{R}^{\geq 1}$. (Although using $K$ for a real number clashes with the notation used throughout the rest of this book, it is standard in the area.) One important convention will be important to keep in mind: for $X$ a subset of $G$ and $n \in \mathbb{N}$, we set $X^n := \{x_1 \cdots x_n : x_1, \ldots, x_n \in X\}$ (so $X^n$ does *not* mean the $n$-fold Cartesian power of $X$).

By a *symmetric* subset of $G$, we mean a set that contains the identity of $G$ and is closed under taking inverse.

**Definition 17.1** $X \subseteq G$ is a *K-approximate group* if $X$ is symmetric and $X^2$ can be covered by at most $K$ left translates of $X$, that is, there are $g_1, \ldots, g_m \in G$ with $m \leq K$ such that $X^2 \subseteq \bigcup_{i=1}^m g_i X$.

*Example 17.2*

1. A 1-approximate subgroup of $G$ is simply a subgroup of $G$.
2. If $X \subseteq G$ is finite, then $X$ is a $|X|$-approximate subgroup of $G$.

The second example highlights that, in order to try to study the general structure of finite $K$-approximate groups, one should think of $K$ as fixed and "small" and

© Springer Nature Switzerland AG 2019

M. Di Nasso et al., *Nonstandard Methods in Ramsey Theory
and Combinatorial Number Theory*, Lecture Notes in Mathematics 2239,
https://doi.org/10.1007/978-3-030-17956-4_17

then try to classify the finite $K$-approximate groups $X$, where $X$ has cardinality much larger than $K$.

**Exercise 17.3** Suppose that $(G, +)$ is an abelian group. For distinct $v_1 \ldots, v_r \in G$ and (not necessarily distinct) $N_1, \ldots, N_r \in \mathbb{N}$, set

$$P(\mathbf{v}, \mathbf{N}) := \{a_1 v_1 + \cdots + a_r v_r \ : \ a_i \in \mathbb{Z}, \ |a_i| \leq N_i\}.$$

Show that $P(\mathbf{v}, \mathbf{N})$ is a $2^r$-approximate subgroup of $G$.

The approximate subgroups appearing in the previous exercise are called *symmetric generalized arithmetic progressions* and the number $r$ of generators is called the *rank* of the progression. The *Freiman Theorem for abelian groups* (due to due to Freiman [51] for $\mathbb{Z}$ and to Green and Ruzsa [62] for a general abelian group) says that approximate subgroups of abelian groups are "controlled" by symmetric generalized arithmetic progressions:

**Theorem 17.4** *There are constants $r_K$, $C_K$ such that the following hold: Suppose that $G$ is an abelian group and $A \subseteq G$ is a finite $K$-approximate group. Then there is a finite subgroup $H$ of $G$ and a symmetric generalized arithmetic progression $P \subseteq G/H$ such that $P$ has rank at most $r_K$, $\pi^{-1}(P) \subseteq \Sigma_4(A)$, and $|P| \geq C_K \cdot \frac{|A|}{|H|}$.*

Here, $\pi \ : \ G \ \to \ G/H$ is the quotient map. For a while it was an open question as to whether there was a version of the Freiman theorem that held for finite approximate subgroups of arbitrary groups. Following a breakthrough by Hrushovski [75], Breuillard et al. [20] were able to prove the following general structure theorem for approximate groups.

**Theorem 17.5** *There are constants $r_K, s_K, C_K$ such that the following hold: Suppose that $G$ is a group and $A \subseteq G$ is a finite $K$-approximate group. Then there is a finite subgroup $H \subseteq G$, a noncommutative progression of rank at most $r_K$ whose generators generate a nilpotent group of step at most $s_K$ such that $\pi^{-1}(P) \subseteq A^4$ and $|P| \geq C_K \cdot \frac{|A|}{|H|}$.*

Here, $\pi : G \to G/H$ is once again the quotient map. To understand this theorem, we should explain the notion of noncommutative progression.

Suppose that $G$ is a group, $v_1, \ldots, v_r \in G$ are distinct, and $N_1, \ldots, N_r > 0$ are (not necessarily distinct) natural numbers. The noncommutative progression generated by $v_1, \ldots, v_r$ with dimensions $N_1, \ldots, N_r$ is the set of words on the alphabet $\{v_1, v_1^{-1}, \ldots, v_r, v_r^{-1}\}$ such that the total number of occurrences of $v_i$ and $v_i^{-1}$ is at most $N_i$ for each $i = 1, \ldots, r$; as before, $r$ is called the rank of the progression. In general, noncommutative progressions need not be approximate groups (think free groups). However, if $v_1, \ldots, v_r$ generate a nilpotent subgroup of $G$ of step $s$, then for $N_1, \ldots, N_r$ sufficiently large, the noncommutative progression is in fact a $K$-approximate group for $K$ depending only on $r$ and $s$. (See, for example, [123, Chapter 12].)

## 17.2   A Special Case: Approximate Groups of Finite Exponent

To illustrate some of the main ideas of the proof of the Breuillard-Green-Tao theorem, we prove a special case due to Hrushovski [75]:

**Theorem 17.6** *Suppose that $X \subseteq G$ is a finite $K$-approximate group. Assume that $X^2$ has exponent $e$, that is, for every $x \in X^2$, we have $x^e = 1$. Then $X^4$ contains a subgroup $H$ of $\langle X \rangle$ such that $X$ can be covered by $L$ left cosets of $H$, where $L$ is a constant depending only on $K$ and $e$.*

Here, $\langle X \rangle$ denotes the subgroup of $G$ generated by $X$. Surprisingly, this theorem follows from the simple observation that the only connected Lie group which has an identity neighborhood of finite exponent is the trivial Lie group consisting of a single point. But how do continuous objects such as Lie groups arise in proving a theorem about finite objects like finite approximate groups? The key insight of Hrushovski is that ultraproducts of finite $K$-approximate groups are naturally "modeled" in a precise sense by second countable, locally compact groups and that, using a classical theorem of Yamabe, this model can be perturbed to a Lie model.

More precisely, for each $i \in \mathbb{N}$, suppose that $X_i \subseteq G_i$ is a finite $K$-approximate group. We set $X := \prod_{\mathscr{U}} X_i$, which, by transfer, is a hyperfinite $K$-approximate subgroup of $G := \prod_{\mathscr{U}} G_i$. In the rest of this chapter, unless specified otherwise, $X$ and $G$ will denote these aforementioned ultraproducts. By a *monadic* subset of $G$ we mean a countable intersection of internal subsets of $G$. Also, $\langle X \rangle$ denotes the subgroup of $G$ generated by $X$.

**Theorem 17.7** *There is a monadic subset $o(X)$ of $X^4$ such that $o(X)$ is a normal subgroup of $\langle X \rangle$ such that the quotient $\mathscr{G} := \langle X \rangle / o(X)$ has the structure of a second countable, locally compact group. Moreover, letting $\pi : \langle X \rangle \to \mathscr{G}$ denote the quotient map, we have:*

1. *The quotient $\langle X \rangle / o(X)$ is bounded, meaning that for all internal sets $A, B \subseteq \langle X \rangle$ with $o(X) \subseteq A$, finitely many left translates of $A$ cover $B$.*
2. *$Y \subseteq \mathscr{G}$ is compact if and only if $\pi^{-1}(Y)$ is monadic; in particular, $\pi(X)$ is compact.*
3. *If $Y \subseteq G$ is internal and contains $o(X)$, then $Y$ contains $\pi^{-1}(U)$ for some open neighborhood of the identity in $\mathscr{G}$.*
4. *$\pi(X^2)$ is a compact neighborhoods of the identity in $\mathscr{G}$.*

Let us momentarily assume that Theorem 17.7 holds and see how it is used to prove Theorem 17.6. As usual, we first prove a nonstandard version of the desired result.

**Theorem 17.8** *Suppose that $X \subseteq G$ is a hyperfinite $K$-approximate group such that $X^2$ has exponent $e$. Then $X^4$ contains an internal subgroup $H$ of $G$ such that $o(X) \subseteq H$.*

*Proof* Let $U$ be an open neighborhood of the identity in $\mathscr{G}$ with $\pi^{-1}(U) \subseteq X^4$ such that $U$ is contained in $\pi(X^2)$, whence $U$ has exponent $e$. By the Gleason-Yamabe theorem [132], there is an open subgroup $\mathscr{G}'$ of $\mathscr{G}$ and normal $N \trianglelefteq \mathscr{G}'$ with $N \subseteq U$ such that $\mathscr{H} := \mathscr{G}'/N$ is a connected Lie group. Let $Y := X \cap \pi^{-1}(\mathscr{G}')$ and let $\rho : \langle Y \rangle \to \mathscr{H}$ be the composition of $\pi$ with the quotient map $\mathscr{G}' \to \mathscr{H}$. Since $\mathscr{G}'$ is clopen in $\mathscr{G}$, $\pi^{-1}(\mathscr{G}')$ is both monadic and co-monadic (the complement of a monadic, also known as *galactic*), whence internal by saturation; it follows that $Y$ is also internal. Since the image of $U \cap \mathscr{G}'$ in $\mathscr{H}$ is also open, it follows that $\mathscr{H}$ is a connected Lie group with an identity neighborhood of finite exponent. We conclude that $\mathscr{H}$ is trivial, whence $\ker(\rho) = Y = \langle Y \rangle$ is the desired internal subgroup of $G$ contained in $X^4$.

*Remark 17.9* The passage from $\mathscr{G}$ to the Lie subquotient $\mathscr{G}'/N$ is called the *Hrushovski Lie Model Theorem*. More precisely, [20] abstracts the important properties of the quotient map $\pi : \langle X \rangle \to \mathscr{G}$ and calls any group morphism onto a second countable, locally compact group satisfying these properties a *good model*. In the proof of Theorem 17.8, we actually showed that the good model $\pi : \langle X \rangle \to \mathscr{G}$ can be replaced by a good model $\rho : \langle Y \rangle \to \mathscr{H}$ onto a connected Lie group. One can show that $Y$ is also an approximate group (in fact, it is a $K^6$-approximate group) that is closely related to the original approximate group $X$, whence the Hrushovski Lie model theorem allows one to study ultraproducts of $K$-approximate groups by working with the connected Lie groups that model them. For example, the proof of Theorem 17.4 actually proceeds by induction on the dimension of the corresponding Lie model. To be fair, the proof of Theorem 17.4 actually requires the use of *local Lie groups* and, in particular, uses the local version of Yamabe's theorem, whose first proof used nonstandard analysis [54].

*Proof of Theorem 17.6* Suppose, towards a contradiction, that the theorem is false. For each $L$, let $G_L$ be a group and $X_L \subseteq G_L$ a finite $K$-approximate group such that $X_L^2$ has exponent $e$ and yet, for any finite subgroup $H$ of $\langle X_L \rangle$ contained in $X_L^4$, we have that $X_L$ is not covered by $L$ cosets of $H$. Let $X := \prod_{\mathscr{U}} X_L$ and $G := \prod_{\mathscr{U}} G_L$. By transfer, $X$ is a $K$-approximate subgroup of $G$ such that $X^2$ has exponent $e$. By Theorem 17.8, $X^4$ contains an internal subgroup $H \supseteq o(X)$ of $\langle X \rangle$. Without loss of generality, we may write $H := \prod_{\mathscr{U}} H_L$ with $H_L$ a subgroup of $G_L$ contained in $X_L^4$. Since the quotient is bounded by Theorem 17.7, there is $M \in \mathbb{N}$ such that $M$ left translates of $H$ cover $X^4$. Thus, for $\mathscr{U}$-almost all $L$, $M$ left translates of $H_L$ cover $X_L$; taking $L > M$ yields the desired contradiction.

We now turn to the proof of Theorem 17.7. Hrushovski's original proof used some fairly sophisticated model theory. A key insight of Breuillard-Green-Tao was that a proof that relied only on fairly elementary combinatorics and nonstandard methods could be given. The following result is the combinatorial core of their proof. It, and the easy lemma after it, do not follow the convention that $X$ is a hyperfinite $K$-approximate group; proofs of both of these results can be found, for example, in [128].

**Theorem 17.10 (Sanders-Croot-Sisask)** *Given $K$ and $\delta > 0$, there is $\epsilon > 0$ so that the following holds: Suppose that $X$ is a finite $K$-approximate subgroup of $G$. Suppose that $Y \subseteq X$ is symmetric and $|Y| \geq \delta|X|$. Then there is a symmetric $E \subseteq G$ such that $|E| \geq \epsilon|X|$ and $(E^{16})^X \subseteq Y^4$.*

**Lemma 17.11** *Let $X \subseteq G$ be a finite $K$-approximate group and $S \subseteq G$ symmetric such that $S^4 \subseteq X^4$ and $|S| \geq c|X|$ for some $c > 0$. Then $X^4$ can be covered by $K^7/c$ left cosets of $S^2$.*

We now return to our assumption that $X$ is a hyperfinite $K$-approximate subgroup of $G$.

**Proposition 17.12** *There is a descending sequence*

$$X^4 =: X_0 \supseteq X_1 \supseteq X_2 \supseteq \cdots \supseteq X_n \supseteq \cdots$$

*of internal, symmetric subsets of $G$ such that:*

(i) $X_{n+1}^2 \subseteq X_n$;

(ii) $X_{n+1}^X \subseteq X_n$;

(iii) $X^4$ *is covered by finitely many left cosets of* $X_n$.

*Proof* Suppose that $Y \subseteq G$ is internal, symmetric, $Y^4 \subseteq X^4$, and $X^4$ can be covered by finitely many left cosets of $Y$. We define a new set $\tilde{Y}$ with these same properties. First, take $\delta > 0$ such that $|Y| \geq \delta|X^4|$; such $\delta$ exists since $X^4$ can be covered by finitely many left cosets of $Y$. By the transfer of Theorem 17.10, there is an internal, symmetric $S \subseteq Y^4$ such that $|S| \geq \epsilon|X^4|$ and $(S^{16})^X \subseteq Y^4$. Let $\tilde{Y} := S^2$. Note that $\tilde{Y}$ has the desired properties, the last of which follows from the preceding lemma.

We now define a sequence $Y_0, Y_1, Y_2, \ldots$, of internal subsets of $X^4$ satisfying the above properties by setting $Y_0 := X$ and $Y_{n+1} := \tilde{Y}_n$. Finally, setting $X_n := Y_n^4$ yields the desired sequence. $\qquad \square$

*Proof of Theorem 17.7* Take $(X_n)$ as guaranteed by Proposition 17.12. We set $o(X) := \bigcap_n X_n$, a monadic subset of $X^4$. It is clear from (i) and (ii) that $o(X)$ is a normal subgroup of $\langle X \rangle$. We can topologize $\langle X \rangle$ by declaring, for $a \in \langle X \rangle$, $\{aX_n : n \in \mathbb{N}\}$ to be a neighborhood base for $a$. The resulting space is not Hausdorff, but it is clear that the quotient space $\langle X \rangle / o(X)$ is precisely the separation of $\langle X \rangle$. It is straightforward to check that the resulting space is separable and yields a group topology on $\mathcal{G}$. Now one uses the boundedness property (proven in the next paragraph) to show that $\mathcal{G}$ is locally compact; see [128] for details.

To show that it is bounded, suppose that $A, B \subseteq \langle X \rangle$ are such that $o(X) \subseteq A$. We need finitely many left cosets of $A$ to cover $B$. Take $n$ such that $X_n \subseteq A$ and take $m$ such that $B \subseteq (X^4)^m$. Since $X^4$ is a $K^4$-approximate group, $(X^4)^m \subseteq E \cdot X^4$ for some finite $E$. By (iii), we have that $X^4 \subseteq F \cdot X_n$ for some finite $F$. It follows that $B \subseteq EFA$, as desired.

The proof that $Y \subseteq \mathcal{G}$ is compact if and only if $\pi^{-1}(Y)$ is monadic is an exercise left to the reader (or, once again, one can consult [128]). To see the moreover part, note that

$$\pi^{-1}(\pi(X)) = \{x \in \langle X \rangle \; : \; \text{there is } y \in X \text{ such that } x^{-1}y \in \bigcap_n X_n\}.$$

In particular, $\pi^{-1}(\pi(X)) \subseteq X^5$ and, by saturation, we actually have

$$\pi^{-1}(\pi(X)) = \{x \in X^5 \; : \; \text{for all } n \text{ there is } y \in X \text{ such that } x^{-1}y \in X_n\}.$$

From this description of $\pi^{-1}(\pi(X))$, we see that it is monadic, whence $\pi(X)$ is compact.

To prove (3), suppose that $Y$ is an internal subset of $G$ containing $o(X)$. Take $n$ such that $X_n \subseteq Y$. Thus, $\pi^{-1}(\pi(X_{n+1})) \subseteq X_n \subseteq Y$ and $\pi(X_{n+1})$ is open in $\mathcal{G}$.

Finally, to see that $\pi(X^2)$ is a neighborhood of the identity in $\mathcal{G}$, first observe that since $X^4$ is covered by finitely many left cosets of $X$, the neighborhood $\pi(X^4)$ of the identity is covered by finitely many left cosets of the compact set $\pi(X)$, whence $\pi(X)$ has nonempty interior and thus $\pi(X^2) = \pi(X) \cdot \pi(X)^{-1}$ is a neighborhood of the identity in $\mathcal{G}$.

**Exercise 17.13** In the notation of the previous proof, show that the space $\mathcal{G}$ is separable and the topology on it is indeed a group topology.

**Exercise 17.14** In the notation of the previous proof, show that $Y \subseteq \mathcal{G}$ is compact if and only if $\pi^{-1}(Y)$ is monadic.

# Notes and References

Although many notions closely related to approximate groups appeared in the literature and were seriously studied, the formal definition of an approximate group appearing above was introduced by Terence Tao in [121] and was in part motivated by its use in the work of Bourgain and Gamburd [19] on superstrong approximation for Zariski dense subgroups of $SL_2(\mathbb{Z})$. The classification of approximate subgroups of $\mathbb{Z}$ was proven by Freiman in [51] and then extended to all abelian groups by Green and Ruzsa [62]. Hrushovski's breakthrough article [75] used sophisticated tools from model theory to make serious progress on the classification of arbitrary approximate groups and these techniques were simplified by Breuillard, Green, and Tao in [20] and extended to complete the classification. While [20] is extremely readable (and includes many interesting and illustrative examples of ultraproducts of finite approximate groups), the Séminaire Bourbaki article of van den Dries [128] provides another thorough treatment, simplifying some of the steps in order to achieve a somewhat weaker conclusion which is still strong enough for many of the combinatorial applications, including a strengthening of Gromov's theorem on groups with polynomial group (whose "standard" proof is nowadays often considered to be the nonstandard one given by van den Dries and Wilkie [129]).

# Appendix A
# Foundations of Nonstandard Analysis

## A.1 Foundations

In this appendix we will revise all the basic notions and principles that we presented in Chap. 2 and put them on firm foundations. As it is customary in the foundations of mathematics, we will work in a set-theoretic framework as formalized by Zermelo-Fraenkel set theory with choice ZFC. Since the purpose of this book is not a foundational one, we will only outline the main arguments, and then give precise bibliographic references where the interested reader can find all proofs worked out in detail.

### A.1.1 Mathematical Universes and Superstructures

Let us start with the notion of a mathematical universe, which formalizes the idea of a sufficiently large collection of mathematical objects that contains all that one needs when applying nonstandard methods.

**Definition A.1** A *universe* $\mathbb{U}$ is a nonempty collection of "mathematical objects" that satisfies the following properties:

1. The numerical sets $\mathbb{N}, \mathbb{Z}, \mathbb{Q}, \mathbb{R}, \mathbb{C} \in \mathbb{U}$;
2. If $a_1, \ldots, a_k \in \mathbb{U}$ then also the tuple $\{a_1, \ldots, a_k\}$ and the ordered tuple $(a_1, \ldots, a_k)$ belong to $\mathbb{U}$;
3. If the family of sets $\mathscr{F} \in \mathbb{U}$ then also its union $\bigcup \mathscr{F} = \bigcup_{F \in \mathscr{F}} F \in \mathbb{U}$;
4. If the sets $A, B \in \mathbb{U}$ then also the *Cartesian product* $A \times B$, the *powerset* $\mathscr{P}(A) = \{A' \mid A' \subseteq A\}$, and the *function set* $\mathrm{Fun}(A, B) = \{f \mid f : A \to B\}$ belong to $\mathbb{U}$;
5. $\mathbb{U}$ is *transitive*, that is, $a \in A \in \mathbb{U} \Rightarrow a \in \mathbb{U}$.

© Springer Nature Switzerland AG 2019
M. Di Nasso et al., *Nonstandard Methods in Ramsey Theory and Combinatorial Number Theory*, Lecture Notes in Mathematics 2239, https://doi.org/10.1007/978-3-030-17956-4

Notice that a universe $\mathbb{U}$ is necessarily closed under subsets; indeed if $A' \subseteq A \in \mathbb{U}$, then $A' \in \mathscr{P}(A) \in \mathbb{U}$, and hence $A' \in \mathbb{U}$, by transitivity. Thus, if the sets $A, B \in \mathbb{U}$ then also the *intersection* $A \cap B$ and the *set-difference* $A \backslash B$ belong to $\mathbb{U}$; moreover, by combining properties 2 and 3, one obtains that also the *union* $A \cup B = \bigcup\{A, B\} \in \mathbb{U}$.

*Remark A.2* It is a well-known fact that all "mathematical objects" used in the ordinary practice of mathematics, including numbers, sets, functions, relations, ordered tuples, and Cartesian products, can all be coded as sets. Recall that, in ZFC, an ordered pair $(a, b)$ is defined as the *Kuratowski pair* $\{\{a\}, \{a, b\}\}$; in fact, it is easily shown that by adopting that definition one has the characterizing property that $(a, b) = (a', b')$ if and only if $a = a'$ and $b = b'$. Ordered tuples are defined inductively by letting $(a_1, \ldots, a_k, a_{k+1}) = ((a_1, \ldots, a_k), a_{k+1})$. A binary relation $R$ is defined as a set of ordered pairs; so, the notion of a relation is identified with the set of pairs that satisfy it. A function $f$ is a relation such that every element $a$ in the domain is in relation with a unique element $b$ of the range, denoted $b = f(a)$; so, the notion of a function is identified with its graph. As for numbers, the natural numbers $\mathbb{N}_0$ of ZFC are defined as the set of *von Neumann naturals*: $0 = \emptyset$ and, recursively, $n + 1 = n \cup \{n\}$, so that each natural number $n = \{0, 1, \ldots, n - 1\}$ is identified with the set of its predecessors; the integers $\mathbb{Z}$ are then defined as a suitable quotient of $\mathbb{N} \times \mathbb{N}$, and the rationals $\mathbb{Q}$ as a suitable quotient of $\mathbb{Z} \times \mathbb{Z}$; the real numbers $\mathbb{R}$ are defined as suitable sets of rational numbers, namely the *Dedekind cuts*; the complex numbers $\mathbb{C} = \mathbb{R} \times \mathbb{R}$ are defined as ordered pairs of real numbers, where the pair $(a, b)$ is denoted $a + ib$. (See, e.g., [74].)

We remark that the above definitions are instrumental if one works within axiomatic set theory, where all notions must be reduced to the sole notion of a set; however, in the ordinary practice of mathematics, one can safely take the ordered tuples, the relations, the functions, and the natural numbers as primitive objects of a different nature with respect to sets.

For convenience, in the following we will consider *atoms*, that is, primitive objects that are not sets.[1] A notion of a universe that is convenient to our purposes is the following.

**Definition A.3** Let $X$ be a set of atoms. The *superstructure* over $X$ is the union $\mathbb{V}(X) := \bigcup_{n \in \mathbb{N}_0} V_n(X)$, where $V_0(X) = X$, and, recursively, $V_{n+1}(X) = V_n(X) \cup \mathscr{P}(V_n(X))$.

**Proposition A.4** *Let $X$ be a set of atoms that includes (a copy of) $\mathbb{N}$. Then the superstructure $\mathbb{V}(X)$ is a universe in the sense of Definition A.1.*[2]

---

[1]The existence of atoms is disproved by the axioms of ZFC, where all existing objects are sets; however, axiomatic theories are easily formalized that allow a proper class of atoms. For instance, one can consider a suitably modified versions of ZFC where a unary predicate $A(x)$ for "$x$ is an atom" is added to the language, and where the axiom of extensionality is restricted to non-atoms.

[2]Clearly, the transitivity property "$a \in A \in \mathbb{V}(X) \Rightarrow a \in \mathbb{V}(X)$" applies provided $A \notin X$.

*Proof* See [25, §4.4].

*Remark A.5* In set theory, one considers the universe $\mathbf{V} = \bigcup_\gamma V_\gamma$ given by the union of all levels of the so-called *von Neumann cumulative hierarchy*, which is defined by transfinite recursion on the class of all ordinals by letting $V_0 = \emptyset$, $V_{\gamma+1} = \mathscr{P}(V_\gamma)$, and $V_\lambda = \bigcup_{\gamma<\lambda} V_\gamma$ if $\lambda$ is a limit ordinal. Basically, the Regularity axiom was introduced in set theory to show that the above class $\mathbf{V}$ is the universal class of all sets.

Instead, the superstructures are defined by only taking the finite levels $V_n(X)$ constructed over a given set of atoms $X$. The main motivation for that restriction is that if one goes beyond the finite levels and allows the first infinite ordinal $\omega$ to belong to the domain of the star map, then $^*\omega$ would contain $\in$-descending chains $\xi \ni \xi - 1 \ni \xi - 2 \ni \ldots$ for every $\xi \in {}^*\omega\backslash\omega$, contradicting the Regularity axiom. Since $V_\omega = \bigcup_{n\in\omega} V_n$ would not be suitable, as it only contains finite sets, one takes an infinite set of atoms $X$ as the ground level $V_0(X) = X$, so as to enclose (a copy of the) natural numbers in the universe.

However, we remark that if one drops the Regularity Axiom from the axioms of ZFC, and replace it with a suitable Anti-Foundation Axiom (such as Boffa's *superuniversality axiom*), then one can construct star maps $* : \mathbf{V} \to \mathbf{V}$ from the universe all sets into itself that satisfies the *transfer principle* and $\kappa$-saturation for any given cardinal $\kappa$. (This is to be contrasted with the well-known result by Kunen about the impossibility in ZFC of non-trivial elementary extensions $j : \mathbf{V} \to \mathbf{V}$.) This kind of foundational issues are the subject matter of the so-called *nonstandard set theory* (see Remark A.15).

## A.1.2  Bounded Quantifier Formulas

In this section we formalize the notion of "elementary property" by means of suitable formulas. It is a well-known fact that virtually all properties of mathematical objects can be described within first-order logic; in particular, one can reduce to the language of set theory grounded on the usual logic symbols plus the sole membership relation symbol. Here is the "alphabet" of our language.[3]

- *Variables*: $x, y, z, \ldots, x_1, x_2, \ldots$;
- *Logical Connectives*: $\neg$ (negation "not"); $\wedge$ (conjunction "and"); $\vee$ (disjunction "or"); $\Rightarrow$ (implication "if ... then"); $\Leftrightarrow$ (double implication "if and only if");
- *Quantifiers*: $\exists$ (existential quantifier "there exists"); $\forall$ (universal quantifier "for all");
- *Equality symbol* $=$;
- *Membership symbol* $\in$.

---

[3]To be precise, also parentheses "(" and ")" should be included among the symbols of our alphabet.

**Definition A.6** An *elementary formula* $\sigma$ is a finite string of symbols in the above alphabet where it is specified a set of *free variables* $FV(\sigma)$ and a set of *bound variables* $BV(\sigma)$, according to the following rules.

* *Atomic formulas.* If $x$ and $y$ are variables then "$(x = y)$" and "$(x \in y)$" are elementary formulas, named *atomic formulas*, where $FV(x = y) = FV(x \in y) = \{x, y\}$ and $BV(x = y) = BV(x \in y) = \emptyset$;
* *Restricted quantifiers.* If $\sigma$ is an elementary formula, $x \in FV(\sigma)$ and $y \notin BV(\sigma)$, then "$(\forall x \in y)\, \sigma$" is an elementary formula where $FV((\forall x \in y)\, \sigma) = (FV(\sigma)\backslash\{x\}) \cup \{y\}$ and $BV((\forall x \in y)\, \sigma) = BV(\sigma) \cup \{y\}$. Similarly with the elementary formula "$(\exists x \in y)\, \sigma$" obtained by applying the existential quantifier;
* *Negation.* If $\sigma$ is an elementary formula then $(\neg \sigma)$ is an elementary formula where $FV(\neg\sigma) = FV(\sigma)$ and $BV(\neg\sigma) = BV(\sigma)$;
* *Binary connectives.* If $\sigma$ and $\tau$ are elementary formulas where $FV(\sigma) \cap BV(\tau) = FV(\tau) \cap BV(\sigma) = \emptyset$, then "$(\sigma \wedge \tau)$" is an elementary formula where $FV(\sigma \wedge \tau) = FV(\sigma) \cup FV(\tau)$ and $BV(\sigma \wedge \tau) = BV(\sigma) \cup BV(\tau)$; and similarly with the elementary formulas $(\sigma \vee \tau)$, $(\sigma \Rightarrow \tau)$, and $(\sigma \Leftrightarrow \tau)$ obtained by applying the connectives $\vee$, $\Rightarrow$, and $\Leftrightarrow$, respectively.

According to the above, every elementary formula is built from *atomic formulas* (and this justifies the name "atomic"). in that an arbitrary elementary formula is obtained from atomic formulas by finitely many iterations of restricted quantifiers, negations, and binary connectives, in whatever order. Only quantifiers produces bound variables, and in fact the bound variables are those that are quantified. Notice that a variable can be quantified only if it is free in the given formula, that is, it actually appears and it has been not quantified already.

It is worth stressing that quantifications are only permitted in the *restricted forms* $(\forall x \in y)$ or $(\exists x \in y)$, where the "scope" of the quantified variable $x$ is "restricted" by another variable $y$. To avoid potential ambiguities, we required that the "bounding" variable $y$ does not appear bound itself in the given formula.

As it is customary in the practice, to simplify notation we will adopt natural shorthands. For instance, we will write "$x \neq y$" to mean "$\neg(x = y)$" and "$x \notin y$" to mean "$\neg(x \in y)$"; we will write "$\forall x_1, \ldots, x_k \in y\, \sigma$" to mean "$(\forall x_1 \in y) \ldots (\forall x_k \in y)\, \sigma$", and similarly with existential quantifiers. Moreover, we will use parentheses informally, and omit some of them whenever confusion is unlikely. So, we may write "$\forall x \in y\, \sigma$" instead of "$(\forall x \in y)\, \sigma$; or "$\sigma \wedge \tau$" instead of "$(\sigma \wedge \tau)$"; and so forth.

Another usual agreement is that negation $\neg$ binds more strongly than conjunctions $\wedge$ and disjunctions $\vee$, which in turn bind more strongly than implications $\Rightarrow$ and double implications $\Leftrightarrow$. So, we may write "$\neg\sigma \wedge \tau$" to mean "$((\neg\sigma) \wedge \tau)$"; or "$\neg\sigma \vee \tau \Rightarrow \upsilon$" to mean "$(((\neg\sigma) \vee \tau) \Rightarrow \upsilon)$"; or "$\sigma \Rightarrow \tau \vee \upsilon$" to mean "$(\sigma \Rightarrow (\tau \vee \upsilon))$".

When writing $\sigma(x_1, \ldots, x_k)$ we will mean that $x_1, \ldots, x_k$ are all and only the free variables that appear in the formula $\sigma$. The intuition is that the truth or falsity of a formula depends only on the values given to its *free variables*, whereas *bound variables* can be renamed without changing the meaning of a formula.

**Definition A.7** A property of mathematical objects $A_1, \ldots, A_k$ is *expressed in elementary form* if it is written down by taking an elementary formula $\sigma(x_1, \ldots, x_k)$, and by replacing all occurrences of each free variable $x_i$ by $A_i$. In this case we denote

$$\sigma(A_1, \ldots, A_k),$$

and we will refer to objects $A_1, \ldots, A_k$ as *constants* or *parameters*.[4] By a slight abuse, sometimes we will simply say *elementary property* to mean "property expressed in elementary form".

The motivation of our definition is the well-known fact that virtually all properties considered in mathematics can be formulated in elementary form. Below is a list of examples that include the fundamental ones. As an exercise, the reader can easily write down by him- or herself any other mathematical property that comes to his or her mind, in elementary form.

*Example A.8* Each property is followed by one of its possible expressions in elementary form.[5]

1. "$A \subseteq B$": $(\forall x \in A)(x \in B)$;
2. $C = A \cup B$: $(A \subseteq C) \wedge (B \subseteq C) \wedge (\forall x \in C)(x \in A \vee x \in B)$;
3. $C = A \cap B$: $(C \subseteq A) \wedge (\forall x \in A)(x \in B \leftrightarrow x \in C)$;
4. $C = A \backslash B$: $(C \subseteq A) \wedge (\forall x \in A)(x \in C \leftrightarrow x \notin B)$;
5. $C = \{a_1, \ldots, a_k\}$: $(a_1 \in C) \wedge \ldots \wedge (a_k \in C) \wedge (\forall x \in C)(x = a_1 \vee \ldots \vee x = a_k)$;
6. $\{a_1, \ldots, a_k\} \in C$: $(\exists x \in C)(x = \{a_1, \ldots, a_k\})$;
7. $C = (a, b)$: $C = \{\{a\}, \{a, b\}\}$[6];
8. $C = (a_1, \ldots, a_k)$ with $k \geq 3$: Inductively, $C = ((a_1, \ldots, a_{k-1}), a_k)$;
9. $(a_1, \ldots, a_k) \in C$: $(\exists x \in C)(x = (a_1, \ldots, a_k))$;
10. $C = A_1 \times \ldots \times A_k$: $(\forall x_1 \in A_1) \ldots (\forall x_k \in A_k)((a_1, \ldots, a_k) \in C) \wedge (\forall z \in C)(\exists x_1 \in A_1) \ldots (\exists x_k \in A_k)(z = (x_1, \ldots, x_k))$;
11. $R$ is a $k$-place relation on $A$: $(\forall z \in R)(\exists x_1, \ldots, x_k \in A)(z = (x_1, \ldots, x_k))$;
12. $f : A \to B$: $(f \subseteq A \times B) \wedge (\forall a \in A)(\exists b \in B)((a, b) \in f) \wedge (\forall a, a' \in A)(\forall b \in B)((a, b), (a', b) \in f \Rightarrow a = a')$;
13. $f(a_1, \ldots, a_k) = b$: $((a_1, \ldots, a_k), b) = (a_1, \ldots, a_k, b) \in f$;
14. $x < y$ in $\mathbb{R}$: $(x, y) \in R$, where $R \subset \mathbb{R} \times \mathbb{R}$ is the order relation on $\mathbb{R}$.

It is worth remarking that a same property may be expressed both in an elementary form and in a non-elementary form. The typical examples involve the powerset operation.

---

[4]In order to make sense, it is implicitly assumed that in every quantification $(\forall x \in A_i)$ and $(\exists x \in A_i)$, the object $A_i$ is a *set* (not an atom).

[5]For simplicity, in each item we use short-hands for properties that have been already considered in previous items.

[6]Recall that ordered pairs $(a, b) = \{\{a\}, \{a, b\}\}$ were defined as *Kuratowski pairs*.

*Example A.9* "$\mathscr{P}(A) = B$" is trivially an elementary property of constants $\mathscr{P}(A)$ and $B$, but *cannot* be formulated as an elementary property of constants $A$ and $B$. In fact, while the inclusion "$B \subseteq \mathscr{P}(A)$" is formalized in elementary form by "$(\forall x \in B)(\forall y \in x)(y \in A)$", the other inclusion $\mathscr{P}(A) \subseteq B$ does not admit any elementary formulation with $A$ and $B$ as constants. The point here is that quantifications over subsets "$(\forall x \subseteq A)(x \in B)$" are not allowed by our rules.

## A.1.3  Łoś' Theorem

The ultrapower construction of the hyperreals is naturally extended to the whole superstructure.

**Definition A.10**  Let $\mathscr{U}$ be an ultrafilter on the set of indexes $I$. The *bounded ultrapower* of the superstructure $\mathbb{V}(X)$ modulo $\mathscr{U}$ is the union

$$\mathbb{V}(X)^I_b / \mathscr{U} := \bigcup_n V_n(X)^I / \mathscr{U}$$

where $V_n(X)^I / \mathscr{U} = \{[f] \mid f : I \to V_n(X)\}$ contains the equivalence classes modulo $\mathscr{U}$ of the $I$-sequences $f$ that take values in the finite level $V_n(X)$. The *pseudo-membership relation* $\in_{\mathscr{U}}$ on $\mathbb{V}(X)^I_b / \mathscr{U}$ is defined by setting:

$$[f] \in_{\mathscr{U}} [g] \iff \{i \in I \mid f(i) \in g(i)\} \in \mathscr{U}.$$

So, the *bounded* ultrapower consists of the equivalence classes modulo $\mathscr{U}$ of the "bounded" $I$-sequences (that is, of those sequences $f : I \to \mathbb{V}(X)$ whose range is included in some finite level $V_n(X)$); and the pseudo-membership holds when the actual membership holds pointwise for $\mathscr{U}$-almost all indexes.

In bounded ultrapowers, properties expressed in elementary form can be interpreted in a natural way.

**Definition A.11**  Let $P$ be a property expressed in elementary form with constant parameters in $\mathbb{V}_b(X)^I / \mathscr{U}$. The *satisfaction relation* "$\mathbb{V}(X)^I_b / \mathscr{U} \models P$" (read: "the property $P$ holds in $\mathbb{V}(X)^I_b / \mathscr{U}$") is defined according to the following rules[7]:

- "$\mathbb{V}(X)^I_b / \mathscr{U} \models [f] = [g]$" when $[f] = [g]$, that is, when $\{i \in I \mid f(i) = g(i)\} \in \mathscr{U}$.
- "$\mathbb{V}(X)^I_b / \mathscr{U} \models [f] \in [g]$" when $[f] \in_{\mathscr{U}} [g]$, that is, when $\{i \in I \mid f(i) \in g(i)\} \in \mathscr{U}$.

---

[7]Following classic logic, we agree that the disjunction "or" is *inclusive*, that is, "$A$ or $B$" is always true except when both A and B are false; and the implication "$A \Rightarrow B$" is true except when $A$ is true and $B$ is false.

- "$\mathbb{V}(X)_b^I/\mathcal{U} \models \neg P$" when "$\mathbb{V}(X)_b^I/\mathcal{U} \not\models P$."
- "$\mathbb{V}(X)_b^I/\mathcal{U} \models (P_1 \wedge P_2)$" when both "$\mathbb{V}(X)_b^I/\mathcal{U} \models P_1$" and "$\mathbb{V}(X)_b^I/\mathcal{U} \models P_2$."
- "$\mathbb{V}(X)_b^I/\mathcal{U} \models (P_1 \vee P_2)$" when "$\mathbb{V}(X)_b^I/\mathcal{U} \models P_1$" or "$\mathbb{V}(X)_b^I/\mathcal{U} \models P_2$."
- "$\mathbb{V}(X)_b^I/\mathcal{U} \models (P_1 \Rightarrow P_2)$" when if "$\mathbb{V}(X)_b^I/\mathcal{U} \models P_1$" then also "$\mathbb{V}(X)_b^I/\mathcal{U} \models P_2$."
- "$\mathbb{V}(X)_b^I/\mathcal{U} \models (P_1 \Leftrightarrow P_2)$" when both "$\mathbb{V}(X)_b^I/\mathcal{U} \models P_1 \Rightarrow P_2$" and "$\mathbb{V}(X)_b^I/\mathcal{U} \models P_2 \Rightarrow P_1$."
- "$\mathbb{V}(X)_b^I/\mathcal{U} \models (\exists x \in [g]) \, \sigma(x, [f_1], \ldots, [f_n])$" when "$\mathbb{V}(X)_b^I/\mathcal{U} \models \sigma([h], [f_1], \ldots, [f_n])$" for some $[h] \in_{\mathcal{U}} [g]$.
- "$\mathbb{V}(X)_b^I/\mathcal{U} \models (\forall x \in [g]) \, \sigma(x, [f_1], \ldots, [f_n])$" when "$\mathbb{V}(X)_b^I/\mathcal{U} \models \sigma([h], [f_1], \ldots, [f_n])$" for every $[h] \in_{\mathcal{U}} [g]$.

Łos' Theorem is a fundamental result in model theory stating that an ultrapower satisfies the same elementary properties as the initial structure. In the case of bounded ultrapowers of superstructures, one has the following formulation.

**Theorem A.12 (Łos)** *Let $\mathbb{V}(X)_b^I/\mathcal{U}$ be a bounded ultrapower of the superstructure $\mathbb{V}(X)$ and let $\sigma([f_1], \ldots, [f_n])$ be a property expressed in elementary form with constant parameters from $\mathbb{V}(X)_b^I/\mathcal{U}$. Then*

$$\mathbb{V}(X)_b^I/\mathcal{U} \models \sigma([f_1], \ldots, [f_n]) \iff \{i \in I \mid \sigma(f_1(i), \ldots, f_n(i)) \text{ holds}\} \in \mathcal{U}.$$

**Corollary A.13** *Let $d : \mathbb{V}(X) \to \mathbb{V}(X)_b^I/\mathcal{U}$ be the diagonal embedding $A \mapsto [\langle A \mid i \in I \rangle]$ of a superstructure into its bounded ultrapower. Then for every property $\sigma(A_1, \ldots, A_n)$ expressed in elementary form with constant parameters $A_j \in \mathbb{V}(X)$ one has*

$$\sigma(A_1, \ldots, A_n) \iff \mathbb{V}(X)_b^I/\mathcal{U} \models \sigma(d(A_1), \ldots, d(A_n)).$$

Usually, in nonstandard analysis one considers a superstructure $\mathbb{V}(X)$, named the *standard universe*, takes a bounded ultrapower $\mathbb{V}(X)_b^I/\mathcal{U}$ of it, and then defines an injective map $\pi : \mathbb{V}(X)_b^I/\mathcal{U} \to \mathbb{V}(Y)$, where $\mathbb{V}(Y)$ is a suitable superstructure called the *nonstandard universe*. Such a map $\pi$, called the *Mostowski collapse*, has the important property that it transforms the pseudo-membership $\in_{\mathcal{U}}$ into actual membership, that is, $[f] \in_{\mathcal{U}} [g] \Leftrightarrow \pi([f]) \in \pi([g])$. As a result, the star map $* = \pi \circ d : \mathbb{V}(X) \to \mathbb{V}(Y)$ obtained by composing the diagonal embedding with the Mostowski collapse satisfies the *transfer* principle. All details of the construction can be found in §4.4 of [25].

Triples $\langle *, \mathbb{V}(X), \mathbb{V}(Y) \rangle$ where the map $* : \mathbb{V}(X) \to \mathbb{V}(Y)$ satisfies the *transfer principle* and $^*X = Y$ are called *superstructure models of nonstandard analysis*.[8]

---

[8]Typically, one takes (a copy of) the real numbers $\mathbb{R}$ as $X$.

## *A.1.4   Models That Allow Iterated Hyper-Extensions*

In applications, we needed iterated hyper-extensions, but in the usual superstructure
approach to nonstandard analysis (recalled in the previous section), such extensions
cannot be accommodated directly. To this end, one would need to construct a
different standard universe each time, which contains the previous nonstandard
universe. A neat way to overcome this problem is to consider a superstructure model
of nonstandard analysis $\langle *, \mathbb{V}(X), \mathbb{V}(X) \rangle$ where the standard and the nonstandard
universe coincide. Clearly, in this case a hyper-extension also belongs to the
standard universe, and so one can apply the star map to it.[9]

   In the following we will assume that arbitrarily large sets of atoms are available.[10]

**Theorem A.14** *Let $\kappa$, $\mu$ be infinite cardinals. Then there exist sets of atoms $X_0 \subset X$
of cardinality $|X_0| = |X| = \mu^\kappa$ and star maps $* : \mathbb{V}(X) \to \mathbb{V}(X)$ such that:*

1. *(a copy of) the real numbers $\mathbb{R} \subset Y$;*
2. *$^*x = x$ for every $x \in X_0$, and hence $^*r = r$ for every $r \in \mathbb{R}$;*
3. *$^*X = X$;*
4. *transfer principle. For every bounded quantifier formula $\varphi(x_1, \ldots, x_n)$ and for
   every $a_1, \ldots, a_n \in \mathbb{V}(X)$:*

$$\varphi(a_1, \ldots, a_n) \iff \varphi(^*a_1, \ldots, ^*a_n);$$

5. *The $\kappa^+$-saturation principle holds.*

*Proof* Since $\mu^\kappa \geq \mathfrak{c}$ has at least the size of the *continuum*, we can pick a set of
atoms $X$ of cardinality $\mu^\kappa$ that contains (a copy of) the real numbers $\mathbb{R}$, and such
that the relative complement $X_0 = X \backslash \mathbb{R}$ has cardinality $\mu^\kappa$. For every $x \in \kappa$, let
$\langle x \rangle = \{a \in \mathrm{Fin}(\kappa) \mid x \in a\}$ be the set of all finite parts of $\kappa$ that contains $x$. It is
readily seen that the family $\{\langle x \rangle \mid x \in \kappa\}$ has the *finite intersection property*, and
so it can be extended to an ultrafilter $\mathcal{U}$ on $I = \mathrm{Fin}(\kappa)$. We now inductively define
maps $\Psi_n : V_n(X)^I/\mathcal{U} \to V_n(X)$ as follows.

   Since $\mu^\kappa = |X| \leq |X^I/\mathcal{U}| \leq |X|^{|I|} = (\mu^\kappa)^\kappa = \mu^\kappa$, we have $|X| = |X^I/\mathcal{U}|$
and we can pick a bijection $\Psi_0 : X \to X^I/\mathcal{U}$ with the property that $\Psi_0(x) = [c_x]_\mathcal{U}$
for every $x \in X_0$. At the inductive step, let $f : I \to V_{n+1}(X)$ be given. If $f(i) \in
V_n(X)$ $\mathcal{U}$-a.e., let $\Psi_{n+1}([f]_\mathcal{U}) = \Psi_n([f]_\mathcal{U})$; and if $f(i) \notin V_n(X)$ $\mathcal{U}$-a.e., that is,
if $f(i) \in \mathscr{P}(V_n(X))$ $\mathcal{U}$-a.e., define

$$\Psi_{n+1}([f]_\mathcal{U}) = \{\Psi_n([g]_\mathcal{U}) \mid g(i) \in f(i) \ \mathcal{U}\text{-a.e.}\}.$$

---

[9]We remark that the notion of "iterated hyper-image" does not make sense in Nelson's *Internal Set
Theory* IST, as well as in other axiomatic theories elaborated upon that approach.

[10]We remark that this is just a simplifying assumption; indeed, in ZFC one can easily construct sets
$X$ of arbitrarily large cardinality that behaves like sets of atoms with respect to the corresponding
superstructures $\mathbb{V}(X)$, that is, such that $\emptyset \notin X$ and $x \cap \mathbb{V}(X) = \emptyset$ for every $x \in X$. Such sets are
named *base sets* in [25, §4.4].

By gluing together the above functions $\Psi_n$, we obtain a map $\Psi : \mathbb{V}(X)^I_b/\mathscr{U} \to$ $\mathbb{V}(X)$ from the bounded ultrapower of our superstructure into the superstructure itself. Finally, define the star map $* : \mathbb{V}(X) \to \mathbb{V}(X)$ as the composition $\Psi \circ d$, where $d$ is the diagonal embedding:

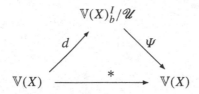

By the definition of $\Psi_0$, for every $x \in X_0$ we have that $*x = \Psi(d(x)) = \Psi_0([c_x]_{\mathscr{U}}) = x$. Moreover, the map $*$ satisfies the *transfer principle* for bounded quantifier formulas, as one can show by using the same arguments as in [25, Theorem 4.4.5]. In brief, the diagonal embedding $d$ preserves the bounded quantifier formulas by Łos' Theorem; moreover it is easily verified from the definition that also $\Psi$ preserves the bounded quantifier formulas. Finally, the range of $\Psi$ is a transitive subset of $\mathbb{V}(X)$, and bounded quantifier formulas are preserved under transitive submodels.

*Remark A.15* The so-called *nonstandard set theories* study suitable adjustments of the usual axiomatic set theory where also the methods of nonstandard analysis are incorporated in their full generality. The most common approach in nonstandard set theories is the so-called *internal viewpoint* as initially proposed independently by Nelson [104] and Hrbacek [73] where one includes in the language a unary relation symbol st for "standard object". The underlying universe is then given by the internal sets, and the standard objects are those internal elements that are hyper-extensions. As a consequence, external sets do not belong to the universe, and can only be considered indirectly, similarly as proper classes are treated in ZFC as extensions of formulas.

An alternative *external viewpoint*, closer to the superstructure approach, is to postulate a suitably modified version of Zermelo-Fraenkel theory ZFC, plus the properties of an *elementary embedding* for a star map $* : \mathbb{S} \to \mathbb{I}$ from the sub-universe $\mathbb{S}$ of "standard" objects into the sub-universe $\mathbb{I}$ of "internal" objects. Of course, to this end one needs to include in the language a new function symbol $*$ for the star map. We remark that if one replaces the *regularity axiom* by a suitable *anti-foundation principle*, then one can actually construct *bounded elementary embeddings* $* : \mathbb{V} \to \mathbb{V}$ defined on the whole universe into itself, thus providing a foundational framework for iterated hyper-extensions that generalizes the superstructure models that we have seen in this section; see [4, 34].

A simple axiomatic presentation to nonstandard analysis that naturally accommodates iterated hyper-extensions is the Alpha-Theory proposed by V. Benci and M. Di Nasso (see the book [10]).

# References

1. S. Albeverio, R. Høegh-Krohn, J.E. Fenstad, T. Lindstrøm, *Nonstandard Methods in Stochastic Analysis and Mathematical Physics*. Pure and Applied Mathematics, vol.122 (Academic Press, Orlando, 1986)
2. R. Arens, The adjoint of a bilinear operation. Proc. Am. Math. Soc. **2**, 839–848 (1951)
3. L.O. Arkeryd, N.J. Cutland, C. Ward Henson (eds.), *Nonstandard Analysis: Theory and Applications*. NATO Advanced Science Institutes Series C: Mathematical and Physical Sciences, vol. 493 (Kluwer Academic Publishers Group, Dordrecht, 1997)
4. D. Ballard, K. Hrbáček, Standard foundations for nonstandard analysis. J. Symb. Log. **57**(2), 741–748 (1992)
5. D. Bartošová, A. Kwiatkowska, Lelek fan from a projective Fraïssé limit. Fund. Math. **231**(1), 57–79 (2015)
6. J.E. Baumgartner, A short proof of Hindman's theorem. J. Comb. Theory. Ser. A **17**, 384–386 (1974)
7. M. Beiglböck, An ultrafilter approach to Jin's theorem. Isr. J. Math. **185**(1), 369–374 (2011)
8. M. Beiglböck, V. Bergelson, A. Fish, Sumset phenomenon in countable amenable groups. Adv. Math. **223**(2), 416–432 (2010)
9. V. Benci, M. Di Nasso, A purely algebraic characterization of the hyperreal numbers. Proc. Am. Math. Soc. **133**(9), 2501–2505 (2005)
10. V. Benci, M. Di Nasso, *How to Measure the Infinite: Mathematics with Infinite and Infinitesimal Numbers* (World Scientific, Singapore, 2018)
11. V. Benci, M. Forti, M. Di Nasso, The eightfold path to nonstandard analysis, in *Nonstandard Methods and Applications in Mathematics*. Lecture Notes in Logic, vol. 25 (Association for Symbolic Logi, La Jolla, 2006), pp. 3–44
12. V. Bergelson, A density statement generalizing Schur's theorem. J. Comb. Theory Ser. A **43**(2), 338–343 (1986)
13. V. Bergelson, Ergodic Ramsey theory—an update, in *Ergodic Theory of $\bf Z^d$ Actions (Warwick, 1993–1994)*. London Mathematical Society Lecture Note Series, vol. 228 (Cambridge University Press, Cambridge, 1996), pp. 1–61
14. V. Bergelson, Ultrafilters, IP sets, dynamics, and combinatorial number theory, in *Ultrafilters Across Mathematics*. Contemporary Mathematics, vol. 530 (American Mathematical Society, Providence, 2010), pp. 23–47
15. V. Bergelson, N. Hindman, Nonmetrizable topological dynamics and Ramsey theory. Trans. Am. Math. Soc. **320**(1), 293–320 (1990)

© Springer Nature Switzerland AG 2019
M. Di Nasso et al., *Nonstandard Methods in Ramsey Theory and Combinatorial Number Theory*, Lecture Notes in Mathematics 2239,
https://doi.org/10.1007/978-3-030-17956-4

16. V. Bergelson, A. Blass, N. Hindman, Partition theorems for spaces of variable words. Proc. Lond. Math. Soc. **68**(3), 449–476 (1994)
17. A. Blass, Selective ultrafilters and homogeneity. Ann. Pure Appl. Logic **38**(3), 215–255 (1988)
18. A. Blass, M. Di Nasso, Finite embeddability of sets and ultrafilters. Bull. Pol. Acad. Sci. Math. **63**(3), 195–206 (2015)
19. J. Bourgain, A. Gamburd, Uniform expansion bounds for Cayley graphs of $SL_2(\mathbb{F}_l)$. Ann. Math. **167**(2), 625–642 (2008)
20. E. Breuillard, B. Green, T. Tao, The structure of approximate groups. Publications mathématiques de l'IHÉS **116**(1), 115–221 (2012)
21. T.C. Brown, An interesting combinatorial method in the theory of locally finite semigroups. Pac. J. Math. **36**, 285–289 (1971)
22. T.C. Brown, V. Rödl, Monochromatic solutions to equations with unit fractions. Bull. Aust. Math. Soc. **43**(3), 387–392 (1991)
23. H. Cartan, Filtres et ultrafiltres. C. R. Acad. Sci. **205**, 777–779 (1937)
24. H. Cartan, Théorie des filtres. C. R. Acad. Sci. **205**, 595–598 (1937)
25. C.C. Chang, H.J. Keisler, *Model Theory*. Studies in Logic and the Foundations of Mathematics, vol. 73, 2nd edn. (North-Holland, Amsterdam, 1977)
26. G. Cherlin, J. Hirschfeld, Ultrafilters and ultraproducts in non-standard analysis, in *Contributions to Non-standard Analysis (Sympos., Oberwolfach, 1970)*. Studies in Logic and the Foundations of Mathematics, vol. 69 (North-Holland, Amsterdam 1972), pp. 261–279
27. P. Cholak, G. Igusa, L. Patey, M. Soskova, D. Turetsky, The rado path decomposition theorem. arXiv:1610.03364 (2016)
28. S. Chow, S. Lindqvist, S. Prendiville, Rado's criterion over squares and higher powers. arXiv:1806.05002 (2018)
29. P. Civin, B. Yood, The second conjugate space of a Banach algebra as an algebra. Pac. J. Math. **11**, 847–870 (1961)
30. W.W. Comfort, S. Negropontis, The theory of ultrafilters. Die Grundlehren der mathematischen Wissenschaften, Band 211, x+482 pp. (Springer, New York, 1974)
31. P. Csikvári, K. Gyarmati, A. Sárközy, Density and Ramsey type results on algebraic equations with restricted solution sets. Combinatorica **32**(4), 425–449 (2012)
32. N.J. Cutland, *Loeb Measures in Practice: Recent Advances*. Lecture Notes in Mathematics, vol. 1751 (Springer, Berlin, 2000)
33. M.M. Day, Amenable semigroups. Ill. J. Math. **1**, 509–544 (1957)
34. M. Di Nasso, An axiomatic presentation of the nonstandard methods in mathematics. J. Symb. Log. **67**(1), 315–325 (2002)
35. M. Di Nasso, An elementary proof of Jin's theorem with a bound. Electron. J. Comb. **21**(2), 2.37 (2014)
36. M. Di Nasso, Embeddability properties of difference sets. Integers **14**, A27 (2014)
37. M. Di Nasso, Hypernatural numbers as ultrafilters, in *Nonstandard Analysis for the Working Mathematician* (Springer, Dordrecht, 2015), pp. 443–474
38. M. Di Nasso, Iterated hyper-extensions and an idempotent ultrafilter proof of Rado's Theorem. Proc. Am. Math. Soc. **143**(4), 1749–1761 (2015)
39. M. Di Nasso, L. Luperi Baglini, Ramsey properties of nonlinear Diophantine equations. Adv. Math. **324**, 84–117 (2018)
40. M. Di Nasso, M. Riggio, Fermat-like equations that are not partition regular. Combinatorica **38**(5), 1067–1078 (2018)
41. M. Di Nasso, I. Goldbring, R. Jin, S. Leth, M. Lupini, K. Mahlburg, High density piecewise syndeticity of sumsets. Adv. Math. **278**, 1–33 (2015)
42. M. Di Nasso, I. Goldbring, R. Jin, S. Leth, M. Lupini, K. Mahlburg, On a sumset conjecture of Erdös. Can. J. Math. **67**(4), 795–809 (2015)
43. M. Di Nasso, I. Goldbring, R. Jin, S. Leth, M. Lupini, K. Mahlburg, High density piecewise syndeticity of product sets in amenable groups. J. Symb. Log. **81**(4), 1555–1562 (2016)

44. R. Ellis, *Lectures on Topological Dynamics* (W. A. Benjamin, New York, 1969)
45. P. Erdős, A survey of problems in combinatorial number theory. Ann. Discrete Math. **6**, 89–115 (1980)
46. P. Erdős, R.L. Graham, *Old and New Problems and Results in Combinatorial Number Theory*. Monographies de L'Enseignement Mathématique, vol. 28 (Université de Genève, L'Enseignement Mathématique, Geneva, 1980)
47. P. Erdős, P. Turán, On some sequences of integers. J. Lond. Math. Soc. **S1-11**(4), 261 (1936)
48. P. Erdös, P. Turán, On a problem of Sidon in additive number theory, and on some related problems. J. Lond. Math. Soc. **16**, 212–215 (1941)
49. E. Følner, Generalization of a theorem of Bogolioùboff to topological abelian groups. With an appendix on Banach mean values in non-abelian groups. Math. Scand. **2**, 5–18 (1954)
50. N. Frantzikinakis, B. Host, Higher order Fourier analysis of multiplicative functions and applications. J. Am. Math. Soc. **30**(1), 67–157 (2017)
51. G.A. Freiman, *Foundations of a Structural Theory of Set Addition*, vol. 37 (American Mathematical Society, Providence, 1973). Translated from the Russian, Translations of Mathematical Monographs
52. H. Furstenberg, Ergodic behavior of diagonal measures and a theorem of Szemerédi on arithmetic progressions. J. Anal. Math. **31**, 204–256 (1977)
53. R. Goldblatt, *Lectures on the Hyperreals*. Graduate Texts in Mathematics, vol. 188 (Springer, New York, 1998)
54. I. Goldbring, Hilbert's fifth problem for local groups. Ann. Math. **172**(2), 1269–1314 (2010)
55. I. Goldbring, S. Leth, On supra-SIM sets of natural numbers. arXiv:1805.05933 (2018)
56. I. Goldbring, H. Towsner, An approximate logic for measures. Isr. J. Math. **199**(2), 867–913 (2014)
57. T. Gowers, Lipschitz functions on classical spaces. Eur. J. Comb. **13**(3), 141–151 (1992)
58. W.T. Gowers, Hypergraph regularity and the multidimensional Szemerédi theorem. Ann. Math. **166**(3), 897–946 (2007)
59. R.L. Graham, B.L. Rothschild, Ramsey's theorem for $n$-parameter sets. Trans. Am. Math. Soc. **159**, 257–292 (1971)
60. R.L. Graham, B.L. Rothschild, A short proof of van der Waerden's theorem on arithmetic progressions. Proc. Am. Math. Soc. **42**, 385–386 (1974)
61. R.L. Graham, K. Leeb, B.L. Rothschild, Ramsey's theorem for a class of categories. Adv. Math. **8**, 417–433 (1972)
62. B. Green, I.Z. Ruzsa, Freiman's theorem in an arbitrary abelian group. J. Lond. Math. Soc. **75**(1), 163–175 (2007)
63. B. Green, T. Sanders, Monochromatic sums and products. Discrete Anal. Paper No. 5, 43 pp. (2016)
64. B. Green, T. Tao, The primes contain arbitrarily long arithmetic progressions. Ann. Math. **167**(2), 481–547 (2008)
65. H. Halberstam, K.F. Roth, *Sequences*, 2nd edn. (Springer, New York, 1983)
66. A.W. Hales, R.I. Jewett, Regularity and positional games. Trans. Am. Math. Soc. **106**(2), 222–229 (1963)
67. D. Hilbert, Über die Irreducibilität ganzer rationaler Functionen mit ganzzahligen Coefficienten. J. Reine Angew. Math. **110**, 104–129 (1892)
68. N. Hindman, The existence of certain ultra-filters on $\mathbb{N}$ and a conjecture of Graham and Rothschild. Proc. Am. Math. Soc. **36**, 341–346 (1972)
69. N. Hindman, Finite sums from sequences within cells of a partition of $\mathbb{N}$. J. Comb. Theory Ser. A **17**, 1–11 (1974)
70. N. Hindman, On density, translates, and pairwise sums of integers. J. Comb. Theory Ser. A **33**(2), 147–157 (1982)
71. N. Hindman, Monochromatic sums equal to products in $\mathbb{N}$. Integers **11**(4), 431–439 (2011)
72. N. Hindman, D. Strauss, *Algebra in the Stone-\v Cech Compactification*. de Gruyter Textbook (Walter de Gruyter, Berlin, 2012)

73. K. Hrbáček, Axiomatic foundations for nonstandard analysis. Polska Akademia Nauk. Fundamenta Mathematicae **98**(1), 1–19 (1978)
74. K. Hrbacek, T. Jech, *Introduction to Set Theory*. Monographs and Textbooks in Pure and Applied Mathematics, vol. 220, 3rd edn. (Marcel Dekker, New York, 1999)
75. E. Hrushovski, Stable group theory and approximate subgroups. J. Am. Math. Soc. **25**(1), 189–243 (2012)
76. R. Jin, Applications of nonstandard analysis in additive number theory. Bull. Symb. Log. **6**(3), 331–341 (2000)
77. R. Jin, Nonstandard methods for upper Banach density problems. J. Number Theory **91**(1), 20–38 (2001)
78. R. Jin, The sumset phenomenon. Proc. Am. Math. Soc. **130**(3), 855–861 (2002)
79. R. Jin, Standardizing nonstandard methods for upper Banach density problems, in *Unusual Applications of Number Theory*. DIMACS: Series in Discrete Mathematics and Theoretical Computer Science, vol. 64 (American Mathematical Society, Providence, 2004), pp. 109–124
80. R. Jin, Introduction of nonstandard methods for number theorists. Integers. Electron. J. Comb. Number Theory **8**(2), A7, 30 (2008)
81. T. Kamae, A simple proof of the ergodic theorem using nonstandard analysis. Isr. J. Math. **42**(4), 284–290 (1982)
82. T. Kamae, M. Keane, A simple proof of the ratio ergodic theorem. Osaka J. Math. **34**(3), 653–657 (1997)
83. H.J. Keisler, An infinitesimal approach to stochastic analysis. Mem. Am. Math. Soc. **48**(297), x+184 (1984)
84. H.J. Keisler, S. Leth, Meager sets on the hyperfinite time line. J. Symb. Log. **56**(1), 71–102 (1991)
85. A. Khalfalah, E. Szemerédi, On the number of monochromatic solutions of $x + y = z^2$. Combin. Probab. Comput. **15**(1–2), 213–227 (2006)
86. E. Lamb, Two-hundred-terabyte maths proof is largest ever. Nature **534**(7605), 17–18 (2016)
87. H. Lefmann, On partition regular systems of equations. J. Comb. Theory Ser. A **58**(1), 35–53 (1991)
88. S. Leth, Some nonstandard methods in combinatorial number theory. Stud. Log. **47**(3), 265–278 (1988)
89. S. Leth, Near arithmetic progressions in sparse sets. Proc. Am. Math. Soc. **134**(6), 1579–1589 (2006)
90. S. Leth, Nonstandard methods and the erdös-turán conjecture, in *The Strength of Nonstandard Analysis* (Springer, Berlin, 2007), pp. 133–142
91. T. Lindstrøm, An invitation to nonstandard analysis, in Nonstandard Analysis and Its Applications (Hull, 1986). London Mathematical Society Student Texts, vol. 10 (Cambridge University Press, Cambridge, 1988), pp. 1–105
92. P.A. Loeb, Conversion from nonstandard to standard measure spaces and applications in probability theory. Trans. Am. Math. Soc. **211**, 113–122 (1975)
93. L. Luperi Baglini, Hyperintegers and nonstandard techniques in combinatorics of numbers. Ph.D. Thesis, University of SIena (2012)
94. L. Luperi Baglini, Partition regularity of nonlinear polynomials: a nonstandard approach. Integers **14**, A30, 23 (2014)
95. L. Luperi Baglini, A nonstandard technique in combinatorial number theory. Eur. J. Comb. **48**, 71–80 (2015)
96. L. Luperi Baglini, $\mathscr{F}$-finite embeddabilities of sets and ultrafilters. Arch. Math. Log. **55**(5–6), 705–734 (2016)
97. M. Lupini, Gowers' Ramsey Theorem for generalized tetris operations. J. Comb. Theory Ser. A **149**, 101–114 (2017)
98. W.A.J. Luxemburg, A general theory of monads, in *Applications of Model Theory to Algebra, Analysis, and Probability (Internat. Sympos., Pasadena, Calif., 1967)* (Holt, Rinehart and Winston, New York, 1969), pp. 18–86

99. W.A.J. Luxemburg, *Non-standard Analysis* (Mathematics Department, California Institute of Technology, Pasadena, 1973)
100. A. Maleki, Solving equations in $\beta\mathbb{N}$. Semigroup Forum **61**(3), 373–384 (2000)
101. J. Moreira, Monochromatic sums and products in $\mathbb{N}$. Ann. Math. **185**(3), 1069–1090 (2017)
102. J. Moreira, F.K. Richter, D. Robertson, A proof of the Erdős sumset conjecture. arXiv:1803.00498 (2018)
103. M.B. Nathanson, Sumsets contained in infinite sets of integers. J. Comb. Theory Ser. A **28**(2), 150–155 (1980)
104. E. Nelson, Internal set theory: a new approach to nonstandard analysis. Bull. Am. Math. Soc. **83**(6), 1165–1198 (1977)
105. D. Ojeda-Aristizabal, Finite forms of Gowers' theorem on the oscillation stability of C 0. Combinatorica **37**, 143–155 (2015)
106. H.J. Prömel, *Ramsey Theory for Discrete Structures* (Springer, Cham, 2013)
107. C. Puritz, Ultrafilters and standard functions in non-standard arithmetic. Proc. Lond. Math. Soc. **22**, 705–733 (1971)
108. C.W. Puritz, Skies, constellations and monads. *Contributions to Non-standard Analysis (Sympos., Oberwolfach, 1970)*. Studies in Logic and Foundations of Mathematics, vol. 69 (North Holland, Amsterdam, 1972), pp. 215–243
109. R. Rado, Studien zur Kombinatorik. Math. Z. **36**(1), 424–470 (1933)
110. R. Rado, Monochromatic paths in graphs. Ann. Discrete Math. **3**, 191–194 (1978)
111. F.P. Ramsey, On a problem of formal logic. Proc. Lond. Math. Soc. **S2-30**(1), 264 (1930)
112. A. Robinson, *Non-standard Analysis* (North-Holland, Amsterdam, 1966)
113. I.Z. Ruzsa, E. Szemerédi, Triple systems with no six points carrying three triangles, in *Combinatorics (Proc. Fifth Hungarian Colloq., Keszthely, 1976), Vol. II*. Colloquia Mathematica Societatis János Bolyai, vol. 18 (North-Holland, Amsterdam, 1978), pp. 939–945
114. J. Sahasrabudhe, Exponential patterns in arithmetic Ramsey theory. arXiv:1607.08396 (2016)
115. J. Sahasrabudhe, Monochromatic solutions to systems of exponential equations. J. Comb. Theory Ser. A **158**, 548–559 (2018)
116. I. Schur, Über die kongruenz $x^n + y^m = z^m \pmod{p}$. Jahresber Deutsch. Math. Verein. **25**, 114–117 (1916)
117. J. Solymosi, Elementary additive combinatorics, in *Additive Combinatorics*. CRM Proceedings & Lecture Notes, vol. 43 (American Mathematical Society, Providence, 2007)
118. C.L. Stewart, R. Tijdeman, On infinite-difference sets. Can. J. Math. **31**(5), 897–910 (1979)
119. E. Szemerédi, On sets of integers containing no $k$ elements in arithmetic progression. Acta Arith. **27**, 199–245 (1975)
120. T. Tao, A correspondence principle between (hyper)graph theory and probability theory, and the (hyper)graph removal lemma. J. Anal. Math. **103**, 1–45 (2007)
121. T. Tao, Product set estimates for non-commutative groups. Combinatorica **28**(5), 547–594 (2008)
122. T. Tao, *An Introduction to Measure Theory*. Graduate Studies in Mathematics, vol. 126 (American Mathematical Society, Providence, 2011)
123. T. Tao, *Hilbert's Fifth Problem and Related Topics*. Graduate Studies in Mathematics, vol. 153 (American Mathematical Society, Providence, 2014)
124. S. Todorcevic, *Introduction to Ramsey Spaces*. Annals of Mathematics Studies, vol. 174 (Princeton University Press, Princeton, 2010)
125. H. Towsner, Convergence of diagonal ergodic averages. Ergodic Theory Dynam. Syst. **29**(4), 1309–1326 (2009)
126. T. Trujillo, From abstract alpha-Ramsey theory to abstract ultra-Ramsey theory. arXiv:1601.03831 [math] (2016)
127. K. Tyros, Primitive recursive bounds for the finite version of Gowers' $c_0$ theorem. Mathematika **61**(3), 501–522 (2015)
128. L. van den Dries, Approximate groups. Astérisque **367**, Exp. No. 1077, vii, 79–113 (2015)

129. L. van den Dries, A.J. Wilkie, Gromov's theorem on groups of polynomial growth and elementary logic. J. Algebra **89**(2), 349–374 (1984)
130. B.L. van der Waerden, Beweis einer baudetschen vermutung. Nieuw Arch. Wisk. **15**, 212–216 (1927)
131. S. Vijay, On the discrepancy of quasi-progressions. Electron. J. Comb. **15**(1), Research Paper 104, 14 (2008)
132. H. Yamabe, A generalization of a theorem of Gleason. Ann. Math. **58**, 351–365 (1953)

# Index

© Springer Nature Switzerland AG 2019
M. Di Nasso et al., *Nonstandard Methods in Ramsey Theory
and Combinatorial Number Theory*, Lecture Notes in Mathematics 2239,
https://doi.org/10.1007/978-3-030-17956-4

# LECTURE NOTES IN MATHEMATICS

Editors in Chief: J.-M. Morel, B. Teissier;

**Editorial Policy**

1. Lecture Notes aim to report new developments in all areas of mathematics and their applications – quickly, informally and at a high level. Mathematical texts analysing new developments in modelling and numerical simulation are welcome.

   Manuscripts should be reasonably self-contained and rounded off. Thus they may, and often will, present not only results of the author but also related work by other people. They may be based on specialised lecture courses. Furthermore, the manuscripts should provide sufficient motivation, examples and applications. This clearly distinguishes Lecture Notes from journal articles or technical reports which normally are very concise. Articles intended for a journal but too long to be accepted by most journals, usually do not have this "lecture notes" character. For similar reasons it is unusual for doctoral theses to be accepted for the Lecture Notes series, though habilitation theses may be appropriate.

2. Besides monographs, multi-author manuscripts resulting from SUMMER SCHOOLS or similar INTENSIVE COURSES are welcome, provided their objective was held to present an active mathematical topic to an audience at the beginning or intermediate graduate level (a list of participants should be provided).

   The resulting manuscript should not be just a collection of course notes, but should require advance planning and coordination among the main lecturers. The subject matter should dictate the structure of the book. This structure should be motivated and explained in a scientific introduction, and the notation, references, index and formulation of results should be, if possible, unified by the editors. Each contribution should have an abstract and an introduction referring to the other contributions. In other words, more preparatory work must go into a multi-authored volume than simply assembling a disparate collection of papers, communicated at the event.

3. Manuscripts should be submitted either online at www.editorialmanager.com/lnm to Springer's mathematics editorial in Heidelberg, or electronically to one of the series editors. Authors should be aware that incomplete or insufficiently close-to-final manuscripts almost always result in longer refereeing times and nevertheless unclear referees' recommendations, making further refereeing of a final draft necessary. The strict minimum amount of material that will be considered should include a detailed outline describing the planned contents of each chapter, a bibliography and several sample chapters. Parallel submission of a manuscript to another publisher while under consideration for LNM is not acceptable and can lead to rejection.

4. In general, **monographs** will be sent out to at least 2 external referees for evaluation.

   A final decision to publish can be made only on the basis of the complete manuscript, however a refereeing process leading to a preliminary decision can be based on a pre-final or incomplete manuscript.

   Volume Editors of **multi-author works** are expected to arrange for the refereeing, to the usual scientific standards, of the individual contributions. If the resulting reports can be forwarded to the LNM Editorial Board, this is very helpful. If no reports are forwarded or if other questions remain unclear in respect of homogeneity etc, the series editors may wish to consult external referees for an overall evaluation of the volume.

5. Manuscripts should in general be submitted in English. Final manuscripts should contain at least 100 pages of mathematical text and should always include

   – a table of contents;
   – an informative introduction, with adequate motivation and perhaps some historical remarks: it should be accessible to a reader not intimately familiar with the topic treated;
   – a subject index: as a rule this is genuinely helpful for the reader.
   – For evaluation purposes, manuscripts should be submitted as pdf files.

6. Careful preparation of the manuscripts will help keep production time short besides ensuring satisfactory appearance of the finished book in print and online. After acceptance of the manuscript authors will be asked to prepare the final LaTeX source files (see LaTeX templates online: https://www.springer.com/gb/authors-editors/book-authors-editors/manuscriptpreparation/5636) plus the corresponding pdf- or zipped ps-file. The LaTeX source files are essential for producing the full-text online version of the book, see http://link.springer.com/bookseries/304 for the existing online volumes of LNM). The technical production of a Lecture Notes volume takes approximately 12 weeks. Additional instructions, if necessary, are available on request from lnm@springer.com.

7. Authors receive a total of 30 free copies of their volume and free access to their book on SpringerLink, but no royalties. They are entitled to a discount of 33.3 % on the price of Springer books purchased for their personal use, if ordering directly from Springer.

8. Commitment to publish is made by a *Publishing Agreement*; contributing authors of multiauthor books are requested to sign a *Consent to Publish form*. Springer-Verlag registers the copyright for each volume. Authors are free to reuse material contained in their LNM volumes in later publications: a brief written (or e-mail) request for formal permission is sufficient.

**Addresses:**
Professor Jean-Michel Morel, CMLA, École Normale Supérieure de Cachan, France
E-mail: moreljeanmichel@gmail.com

Professor Bernard Teissier, Equipe Géométrie et Dynamique,
Institut de Mathématiques de Jussieu – Paris Rive Gauche, Paris, France
E-mail: bernard.teissier@imj-prg.fr

Springer: Ute McCrory, Mathematics, Heidelberg, Germany,
E-mail: lnm@springer.com

Printed in the United States
By Bookmasters